CLINICAL

Fundamentals
of Anatomy
and Physiology

CLINICAL MANUAL

second edition

Fundamentals of Anatomy and Physiology

Frederic Martini, Ph.D.

with

William C. Ober, M.D.
Art coordinator and illustrator

Claire W. Garrison, R.N.
Illustrator

Kathleen Welch, M.D.
Clinical consultant

Prentice Hall, Englewood Cliffs, New Jersey 07632

Editorial Production/Supervision
        and Interior Design: **Carolyn Del Corso**
Pre-press Buyer: **Paula Massenaro**
Manufacturing Buyer: **Lori Bulwin**
Acquisitions Editor: **David Kendric Brake**
Development Editor: **Dan Schiller**

©1992 by Prentice-Hall, Inc.
A Simon & Schuster Company
Englewood Cliffs, New Jersey 07632

Printed in the United States of America

10   9   8   7   6   5   4   3   2   1

ISBN 0-13-335357-5

Prentice-Hall International (UK) Limited, *London*
Prentice-Hall of Australia Pty. Limited, *Sydney*
Prentice-Hall Canada Inc. *Toronto*
Prentice-Hall Hispanoamericana, S.A., *Mexico*
Prentice-Hall of India Private Limited, *New Delhi*
Prentice-Hall of Japan, Inc., *Tokyo*
Simon & Schuster Asia Pte. Ltd., *Singapore*
Editora Prentice-Hall do Brasil, Ltda., *Rio de Janeiro*

Contents

Preface

This manual has been written as a companion volume to the second edition of *Fundamentals of Anatomy and Physiology.* It is my firm belief that an introductory text in this field should do more than explain the mechanics of physiological systems. It should also serve as a reference, with practical information that can provide perspective on the health matters that affect the lives of students and their families. To meet these goals the clinical material, must be concise, understandable, and comprehensive enough to avoid errors caused by oversimplification.

The first edition of *Fundamentals of Anatomy and Physiology* contained large numbers of in-text boxes dealing with clinical concepts. Although effective as a reference, this approach had two major drawbacks: the large number of boxes interrupted the visual flow of information, and topics and depth of coverage were often limited by space constraints. The Clinical Manual evolved to address these problems.

The combination of the material in the Clinical Manual and the boxed material in the text provides an introduction to the major pathological conditions and diagnostic procedures encountered in clinical practice. Few instructions are likely to cover all of the material in the Clinical Manual. Indeed, some instructors may not choose to cover all of the boxes in the text. Because courses differ in their focus, and students differ in their abilities and interests, the aim has been to allow for the maximum possible degree of flexibility. The selection of topics in the text boxes and the Clinical Manual will enable the instructor to use topical news items, personal experiences, or classroom questions as jumping-off points for explorations that vividly demonstrate the relevance of the material being studied, as well as the relationship between normal and abnormal physiology. Boxes and Clinical Manual discussions that are not covered in class can still be assigned for study, suggested as interesting reading, drawn on for reference, or left to the motivated student to discover on his or her own initiative. Each student is likely to find and read selections that answer important questions concerning disorders or procedures experienced by friends or family, or that are relevant to a contemplated career path.

The Clinical Manual is organized in chapters that parallel those of the textbook, and the two are clearly cross-referenced. Users of *Fundamentals of Anatomy and Physiology* will find each Clinical Manual discussion identified by its title and distinctive icon, which immediately follow the relevant passage in the text. (In addition, all Clinical Manual discussions pertaining to a particular chapter are listed in the Review Planner at the end of that chapter.) Conversely, all topics in the Clinical Manual are accompanied by page references to the corresponding material in the text. <u>The appropriate section of *Fundamentals of Anatomy and Physiology* should always be reviewed before consulting the Clinical Manual; clinical conditions can be understood only by comparison with normal system function and regulation.</u>

In addition to this clinically oriented material, the Manual includes two additional sections that should prove highly useful to students of anatomy and physiology. The first consists of 34 cadaver dissection photographs, selected from the set of nearly 300 slides available with the text. The second, "Survival and Beyond," discusses a variety of skills, techniques, and habits that can help students to master the material in their text and course.

CHAPTER 2

The Chemical Level of Organization

Solute Concentrations

Page 39

Physiologists and clinicians often monitor solute concentrations in body fluids such as blood or urine. Data may be reported in several different ways, and two methods will be introduced here. The first method is to give the number of atoms, molecules, or ions in a specific volume of solution. Values are reported in terms of **moles** (mol, M) or **millimoles** (mmol, mM) per liter. A mole is defined as the number of atoms or molecules in a sample that has a weight in grams equal to the atomic weight (for an element) or the molecular weight (for a molecule). Expressing relationships in moles rather than grams makes it much easier to keep track of the relative numbers of atoms or molecules involved. Physiological concentrations are often stated in millimoles per liter (mmol/ℓ). Table 2-A indicates the concentrations of some important solutes in the blood.

Concentrations can be reported in terms of millimoles only when dealing with a specific ion or molecule of known molecular weight. When you cannot be very specific about the chemical substance involved, a second method is used. This

■ TABLE 2-A Solute Concentrations in the Blood		
Solute	*mmol/ℓ*	*mg/dℓ*
Inorganic ions		
Sodium (Na)$^+$	140	320
Potassium (K$^+$)	4.2	16,4
Calcium (Ca^{2+})	2.4	9.5
Chloride (Cl$^-$)	100	354
Organic solutes		
Glucose	5	90
Lipids, total	nr	600
Proteins, total	nr	7–9 g/dℓ

Note: For the purposes of this table values were selected from within normal ranges. nr = standard tests do not report results in these units.

method reports the weight of material dissolved in a unit volume of solution. As indicated in Table 2-A, values are most often expressed in terms of grams (g), milligrams (mg), or micrograms (μg) per deciliter (dℓ, or 100 mℓ).

Appendix IV of the text discusses these and other methods of reporting physiological data, and Appendix V contains a detailed analysis of blood and other body fluids.

CHAPTER 3

The Cellular Level of Organization: Cell Structure

 ## Mitochondrial DNA, Disease, and Evolution

Page 70

There are several inheritable disorders that result from abnormal mitochondrial activity. The mitochondria involved have defective enzymes that reduce their ability to generate ATP. Cells throughout the body may be affected, but symptoms involving muscle cells, nerve cells, and the receptor cells in the eye are most commonly seen because these cells have especially high energy demands. Disorders caused by defective mitochondria are called **mitochondrial cytopathies**. In several instances, the disorder results from abnormal mitochondrial DNA. Examples include certain types of epilepsy (**myoclonic epilepsy**) and blindness (**Leber's hereditary optic neuropathy**).

The pattern of inheritance is unusual. Although men or women may have the disease, only women can pass the condition on to their children. The explanation for this pattern is that the disorder results from an abnormality in the DNA of mitochondria, not in the DNA of cell nuclei. All of the mitochondria in the body are produced through the replication of mitochondria present in the fertilized egg. None of those mitochondria were provided by the father; the mitochondria of the sperm do not remain intact after fertilization takes place. As a result, children can inherit these conditions only from their mothers.

This brings us to an interesting concept. All of your mitochondria were inherited from your mother, and hers from her mother, and so on back through time. The same is true for every other human being. Now, it is known that over long periods of time small changes in DNA nucleotide sequences accumulate. Mitochondrial DNA, or *mDNA*, can therefore be used to estimate the degree of relationship between individuals. The greater the difference between the mDNA of two individuals, the more time has passed since the lifetime of their most recent common ancestor, and the more distant their relationship. On this basis it has been estimated that all human beings now alive shared a common female ancestor roughly 350,000 years ago. Appropriately enough, that individual has been called a "Mitochondrial Eve."

Lysosomal Storage Diseases   *Page 76*

Problems with lysosomal enzyme production cause more than 30 **storage diseases** affecting children. In these conditions the lack of a specific lysosomal enzyme results in the buildup of materials normally removed and recycled by lysosomes. Eventually the cell cannot continue to function. One example is **glycogen storage disease** (Type II), which primarily affects skeletal muscle, cardiac muscle, and liver cells—all cells that synthesize and store glycogen. In this condition the cells are unable to mobilize glycogen normally, and large numbers of insoluble glycogen granules accumulate in the cytoplasm. These granules disrupt the organization of the cytoskeleton, interfering with transport operations and the synthesis of materials. In skeletal and heart muscle cells, the buildup leads to muscular weakness and potentially fatal heart problems.

CHAPTER 5
The Tissue Level of Organization

Marfan's Syndrome   *Page 140*

Marfan's syndrome is an inherited condition that results from the abnormal formation of connective tissue. In this disorder, which affects about 1 person in 10,000 in the United States, a genetic defect causes the production of an abnormal form of *fibrillin*, a glycoprotein important to normal connective tissue strength and elasticity.

Because connective tissues are found in most organs, the effects of this defect are widespread. The most visible sign of Marfan's syndrome involves the skeleton; individuals with Marfan's syndrome are usually tall, with abnormally long arms, legs, and fingers. But the most serious consequences involve the cardiovascular system. Roughly 90 percent of the people with Marfan's syndrome have abnormal cardiovascular systems. Inside the heart, connective tissue supports the valves that ensure a one-way flow of blood. In Marfan's syndrome, these valves are often defective, and the heart works at reduced efficiency. Outside the heart, connective tissue that reinforces the walls of the aorta, a large blood vessel leaving the heart, may be too weak to resist the blood pressure. As a result, a bubble forms in the vessel wall. If the bubble breaks, there is a sudden, fatal loss of blood. Marfan's syndrome is discussed further in Chapter 7 of this Manual (see Abnormalities in Skeletal Development).

CHAPTER 6

The Integumentary System

Excessive Keratin Production

Page 186

Excessive production of keratin is called **hyperkeratosis** (hī-per-ker-a-TŌ-sis). The effects are easily observed as *calluses* and *corns*. Calluses appear on thick-skinned areas exposed to mechanical stress, such as the palms of the hands or the heels. Corns are more localized, and form on or between the toes.

In **psoriasis** (so-RĪ-a-sis) the stratum germinativum becomes unusually active in specific areas, including the scalp, elbows, palms, soles, groin, and nails. Normally an individual stem cell divides once every 20 days, but in psoriasis it may divide every day-and-a-half. Keratinization is abnormal and often incomplete by the time the outer layers are shed. The affected areas appear to be covered with small scales that continually flake away. Most cases are painless and treatable.

Psoriasis appears in 20–30 percent of the individuals with an inherited tendency for the condition. Roughly 5 percent of the general U.S. population has psoriasis to some degree, often aggravated by stress and anxiety.

Dermatitis

Page 190

Dermatitis is an inflammation of the skin that primarily involves the papillary region of the dermis. The inflammation usually begins in a portion of the skin exposed to infection or irritated by chemicals, radiation, or mechanical stimuli. Dermatitis may cause no physical discomfort, or it may produce an annoying itch, as in poison ivy. Other forms of this condition can be quite painful, and the inflammation may spread rapidly across the entire integument.

There are many forms of dermatitis, some of them quite common. **Contact dermatitis** usually occurs in response to strong chemical irritants. It produces an itchy rash that may spread to other areas; poison ivy is an example. **Eczema** (EK-se-ma) is a dermatitis that can be triggered by temperature changes, fungus, chemical irritants, greases, detergents, or stress. Hereditary or environmental factors or both can encourage the development of eczema. **Diaper rash** is a localized dermatitis caused by a combination of moisture, irritating chemicals from fecal or urinary wastes, and flourishing microorganisms. **Urticaria** (ur-ti-KAR-ē-a), also known as **hives,** is an extensive allergic response to a food, drugs, an insect bite, infection, stress, or other stimulus.

Acne

Page 191

Individuals with a genetic tendency toward **acne** have larger than average sebaceous glands, and when the ducts become blocked the secretions accumulate. Inflammation develops, and bacterial infection may occur. The condition usually surfaces at puberty, as production of sex hormones accelerates. The secretory output of the sebaceous glands may be further encouraged by anxiety, stress, physical exertion, certain foods, and drugs.

The visible signs of acne are called **comedos** (ko-ME-dōz). "Whiteheads" contain accumulated, stagnant secretions. "Blackheads" contain more solid material that has been invaded by bacteria. Although neither condition indicates the presence of dirt in the pores, washing may help to reduce superficial oiliness.

Acne usually fades after sex hormone concentrations stabilize. Topical (applied) antibiotics, vitamin A derivatives such as *tretinoin (Retin-A)*, or peeling agents may help reduce inflammation and minimize scarring. In cases of severe acne, the most effective treatment usually involves the discouragement of bacteria by the administration of antibiotic drugs. Because oral antibiotic therapy has risks, including the development of antibiotic-resistant bacteria, this therapy is not used unless other treatment methods have failed.

Truly dramatic improvements in severe cases of acne have been obtained with the prescription drug *Accutane*®. This compound is structurally similar to vitamin A, and it reduces sebaceous gland activity on a long-term basis. A number of minor side effects, including skin rashes, have been reported; these apparently disappear when the treatment ends. However, the use of *Accutane* during the first month of pregnancy carries a high (25 times normal) risk of inducing birth defects.

Two factors interact to determine baldness. A bald individual has a genetic susceptibility, but this tendency must be triggered by large quantities of male sex hormones. Many women carry the genetic background for baldness, but unless major hormonal abnormalities develop, as in certain endocrine tumors, nothing happens.

Male pattern baldness affects the top of the head first, only later reducing the hair density along the sides. Thus hair follicles can be removed from the sides and implanted on the top or front of the head, temporarily delaying a receding hairline. This procedure is rather expensive (thousands of dollars), and not every transplant is successful.

Alopecia areata (al-ō-PĒ-shē-ah ar-ē-A-ta) is a localized hair loss that can affect either sex. The cause is not known, and the severity of hair loss varies from case to case. This condition is associated with several disorders of the immune system; it has also been suggested that periods of stress may promote alopecia areata in individuals already genetically prone to baldness.

Skin conditions that affect follicles can contribute to hair loss. Baldness can also result from exposure to radiation or to many of the toxic (poisonous) drugs used in cancer therapy. Both radiation and anticancer drugs have their greatest effects on rapidly dividing cells, and thus tend to damage matrix cells.

Hairs are dead, keratinized structures, and no amount of oiling, shampooing, or dousing with kelp extracts, vitamins, or nutrients will influence the growth of either the exposed hair or the follicle buried in the dermis. Untested treatments for baldness were banned by the Food and Drug Administration in 1984. *Minoxidil*, a drug originally marketed for the control of blood pressure, appears to stimulate hair follicles when rubbed onto the scalp. It is now available on a prescription basis. Treatment involves applying a 2 percent solution to the scalp twice daily; after 4 months, over one-third of patients reported satisfactory results.

Hirsutism (HER-sut-izm; *hirsutus*, bristly) refers to the growth of hair on women in patterns usually characteristic of men. Because considerable overlap exists between the two sexes in terms of normal hair distribution, and there are significant racial and genetic differences, the precise definition is more often a matter of personal taste than objective analysis. Age and sex hormones may play a role, for hairiness increases late in pregnancy, and menopause produces a change in body hair patterns. Severe hirsutism is often associated with abnormal androgen (male sex hormone) production, either in the ovaries or in other endocrine organs.

Unwanted follicles can be permanently "turned off" by plucking a growing hair and removing the papilla. Electrocautery, which destroys the follicle with a jolt of

7

electricity, requires the services of a professional, but the results are more reliable than plucking. In Europe, but not in the United States, patients may be treated with drugs that prevent androgen stimulation of the follicles.

Complications of Inflammation Page 199

When bacteria invade the dermis, the production of baterial toxins and the debris from dead cells powerfully stimulate the inflammation process. *Pus* is an accumulation of debris, fluid, dead and dying cells, and necrotic tissue components. Pus typically forms at the site of an infection in the dermis. When pus accumulates in an enclosed tissue space, the result is an *abscess.* In the skin, an abscess can form as pus builds up inside the fibrin clot that surrounds the injury site. If the cellular defenses succeed in destroying the invaders, the pus will either be absorbed or surrounded by a fibrous capsule, creating a **cyst**.

Erysipelas (er-i-SIP-e-las; *erythros*, red + *pella*, skin) is a widespread inflammation of the dermis caused by bacterial infection. If the inflammation spreads into the subcutaneous layer and deeper tissues, the condition is called **cellulitis** (sel-ē-LĪ-tis). Erysipelas and cellulitis develop when bacterial invaders break through the fibrin wall. The bacteria involved produce large quantities of *hyaluronidase*, an enzyme that liquifies the ground substance, and *fibrinolysin*, an enzyme that breaks down fibrin and prevents clot formation. (The role of hyaluronidase was discussed in Chapter 5 of the text—see the Clinical Comment on p. 140.) These are serious conditions that require prompt antibiotic therapy.

An **ulcer** is a localized shedding of an epithelium. **Decubitus ulcers**, also known as "bedsores," may afflict bedridden or mobile patients with circulatory restrictions, especially when splints, casts, or bedding continually press against superficial blood vessels. Such sores most often affect the skin near joints or projecting bones, where the vessels are pressed against underlying structures. The chronic lack of circulation kills epidermal cells, removing a barrier to bacterial infection, and eventually the dermal tissues deteriorate as well. (A comparable necrosis will occur in any tissues deprived of adequate circulation.) Bedsores can be prevented or treated by frequent changes in body position that vary the pressures applied to specific blood vessels.

A Classification of Wounds Page 199

Injuries, or **trauma**, involving the integument are very common, and a number ofterms are used to describe them. Each type of wound presents a different series of problems to clinicians attempting to limit damage and promote healing.

An **open wound** is an injury that produces a break in the epithelium. The major categories of open wounds are illustrated in Figure 6-A. **Abrasions** are the result of scraping against a solid object. Bleeding may be slight, but a considerable area may be open to invasion by microorganisms. **Incisions** are linear cuts produced by sharp objects. Bleeding may be severe if deep vessels are damaged. The bleeding may help to flush the wound, and closing the incision with bandages or stitches can limit the area open to attack while healing is under way. A **laceration** is a jagged, irregular tear in the surface produced by solid impact or an irregular object. Tissue damage is more extensive than that caused by an incision, and repositioning the opposing sides of the injury may be difficult. Despite the bleeding that usually occurs, lacerations are prone to infection. **Punctures** result from slender, pointed objects piercing the epithelium. Little bleeding results, and any microbes delivered in the process are likely to find conditions to their liking. In an **avulsion**, chunks of tissue are torn away by the brute force of an automobile accident, explosion, or other such incident. Bleeding may be considerable, and even more serious internal damage may be present.

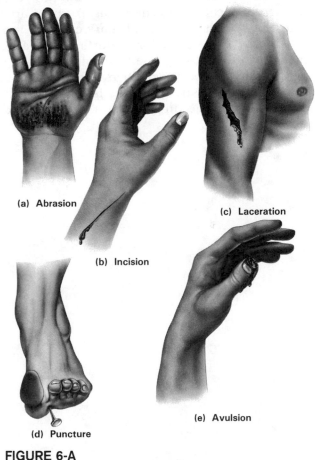

(a) Abrasion

(b) Incision

(c) Laceration

(d) Puncture

(e) Avulsion

FIGURE 6-A
Major types of open wounds

Closed wounds may affect any internal tissue, but because the epithelium is intact the likelihood of infection is reduced. A **contusion** is a bruise that causes bleeding in the dermis. "Black-and-blue" marks are familiar examples of contusions that are taken lightly; contusions of the head, such as "black eyes," may be harmless or a sign of dangerous intracranial bleeding. In general, closed wounds that affect internal organs and organ systems are serious threats to life.

Synthetic Skin   *Page 204*

As indicated on p. 204 of the text, epidermal culturing can produce a new epithelial layer to cover a burn site. A second new procedure provides a model for dermal repairs that takes the place of normal granulation tissue. A special synthetic skin is used. The imitation has a plastic (silastic) "epidermis" and a dermis composed of collagen fibers and ground cartilage. The collagen fibers are taken from cow skin, and the cartilage from sharks. (The use of shark cartilage represents a second contribution to medicine by these predators; the first, antiangiogenesis factor, was noted in Chapter 5 of the text—see the Clinical Comment on p. 147.) Over time, fibroblasts migrate among the collagen fibers and gradually replace the model framework with their own. The silastic epidermis is intended only as a temporary cover that will be replaced by either skin grafts or a cultured epidermal layer.

CHAPTER 7

The Skeletal System: Osseous Tissue and Skeletal Structure

Inherited Abnormalities in Skeletal Development

Page 218

There are several inherited conditions that result in abnormal bone formation. Three examples are *osteogenesis imperfecta, Marfan's syndrome*, and *achondroplasia*.

Osteogenesis imperfecta (im-per-FEK-ta) is an inherited condition, appearing in 1 individual in about 20,000, that affects the organization of collagen fibers. Osteoblast function is impaired, growth is abnormal, and the bones are very fragile, leading to progressive skeletal deformation and repeated fractures. Fibroblast activity is also affected, and the ligaments and tendons may become very "loose," permitting excessive movement at the joints.

Marfan's syndrome is also linked to defective collagen fiber production. This condition, which affects approximately 1 individual in 10,000, was discussed in Chapter 5 of this Manual. Extremely long and slender limbs are the most obvious physical indication of this disorder. The abnormal proportions result from excessive cartilage formation at the epiphyseal plates.

Achondroplasia (ā-kon-drō-PLĀ-sē-a) is another condition resulting from abnormal epiphyseal activity. In this case the epiphyseal plates grow unusually slowly, and the individual develops short, stocky limbs. Although there are other skeletal abnormalities, the trunk is normal in size, and sexual and mental development remain unaffected. The adult will be an **achondroplastic dwarf**.

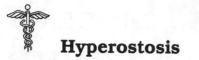

The excessive formation of bone is termed **hyperostosis** (hī-per-os-TŌ-sis). In **osteopetrosis** (os-tē-ō-pe-TRŌ-sis; *petros*, stone) the total mass of the skeleton gradually increases because of a decrease in osteoclast activity. Remodeling stops, and the shapes of the bones gradually change. Osteopetrosis in children produces a variety of skeletal deformities. The primary cause for this relatively rare condition is unknown.

In **acromegaly** (*akron*, extremity + *megale*, great) an excessive amount of growth hormone is released after puberty, when most of the epiphyseal plates have already closed. Cartilages and small bones respond to the hormone, however, resulting in abnormal growth at the hands, feet, lower jaw, skull, and clavicle. Figure 7-A shows a typical acromegalic individual.

FIGURE 7-A
A person suffering from acromegaly

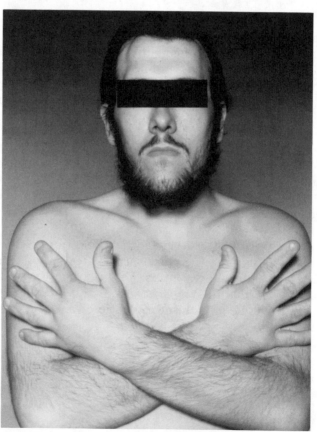

Despite the considerable capacity for bone repair, every fracture does not heal as expected. A **delayed union** is one that proceeds more slowly than anticipated. **Nonunion** may occur as a result of complicating infection, continued movement, or other factors preventing complete callus formation.

There are several techniques for inducing bone repair. Surgical bone grafting is the most common treatment for nonunion. This method immobilizes the bone fragments and provides a bony model for the repair process. Dead bone or bone fragments can be used; alternatively, living bone from another site, such as the fibula (the slender bone of the lower leg) can be inserted. As an alternative to bone grafting, surgeons can insert a shaped patch, made by mixing crushed bone and water.

Another approach involves the stimulation of osteoblast activity by strong electrical fields at the injury site. This procedure has been used to promote bone growth after fractures have refused to heal normally. Wires may be inserted into the skin, implanted in the adjacent bone, or wrapped around a cast. The overall success rate of about 80 percent is truly impressive.

One experimental method of inducing bone repair involves mixing bone marrow cells into a soft matrix of bone collagen and ceramic. This combination is used like a putty at the fracture site. Mesenchymal cells in the marrow divide, producing chondrocytes that create a cartilaginous patch that is later converted to bone by periosteal cells. A second experimental procedure uses a genetically engineered protein to stimulate the conversion of osteoprogenitor cells into active osteoblasts. Although results in animal experimentation have been promising, neither technique has yet been approved for human trials.

Heterotopic (*hetero*, different + *topos*, place), or **ectopic** (*ektos*, outside), bones are those that develop in unusual places. Such bones demonstrate the adaptability of connective tissues quite dramatically. Physical or chemical events can stimulate the development of osteoblasts in normal connective tissues. For example, sesamoid bones develop within tendons near points of friction and pressure. Bone may also appear within a large blood clot at an injury site or within portions of the dermis subjected to chronic abuse. Other triggers include foreign chemicals and problems that affect calcium excretion and storage.

Almost any connective tissues may be affected. Ossification within a tendon or around joints can produce a painful interference with movement. Bone formation may also occur within the kidneys, between skeletal muscles, within the pericardium, in the walls of arteries, or around the eyes.

Myositis ossificans (mī-ō-SĪ-tis os-SIF-i-kans) involves the deposition of bone around skeletal muscles. A muscle injury can trigger a minor case. Severe cases have no known cause, but they certainly provide the most dramatic demonstrations of heterotopic bone formation. If the process does not reverse itself, the muscles

FIGURE 7-B
Myositis ossificans

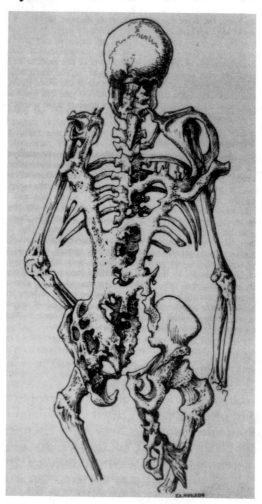

of the back, neck, and upper limbs will gradually be replaced by bone. The extent of the conversion can be seen in Figure 7-B. The drawing shows the skeleton of a 39-year-old man with advanced myositis ossificans. Several of the vertebrae have fused into a solid mass, and major muscles of the back, shoulders, and hips have undergone extensive ossification. The specific muscles ban be identified by comparing this figure with Figures 11-9, 11-13, 11-14, 11-19, and 11-20. (© *1974, L. B. Halstead Wykeman Publications (London) Ltd*)

CHAPTER 8

The Skeletal System: Axial Division

Phrenology   *Page 254*

Phrenology (fre-NOL-o-jē) was a popular "science" of the late eighteenth century. It was argued that since the brain was responsible for intellectual abilities and the skull housed the brain, any bumps and lumps on the skull would reflect features of the brain responsible for the individual's unique, specific traits and abilities. Although this theory fell into disfavor in the early 1800s, it eventually resurfaced a century later as investigators analyzed the size, weight, and superficial features of the brain. Prestigious institutions devoted considerable time and effort to acquiring brain collections, in the hopes of establishing correlations between brain characteristics and intellectual abilities, personality traits, and suitable professions. Their attempts failed, as no such correlation exists. However, the collections continue to make interesting paperweights and educational displays. Brain size and shape appear to be quite variable factors, and as long as growth is permitted, the functional capabilities will not be affected.

15

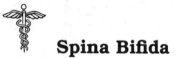

The Thoracic Cage and Surgical Procedures

Surgery on the heart, lungs, or other organs in the thorax often involves entering the thoracic cavity. The mobility of the ribs and the cartilaginous connections with the sternum allow the ribs to be temporarily moved out of the way. Special rib-spreaders are used, which push them apart in much the same way that a jack lifts a car off the ground for a tire change. If more extensive access is required, the sternal cartilages can be cut and the entire sternum can be folded out of the way. Once replaced, the cartilages are reunited by scar tissue, and the ribs heal fairly rapidly.

After thoracic surgery, **chest tubes** may penetrate the thoracic wall to permit drainage of fluids. To install a chest tube or obtain a sample of pleural fluid, the wall of the thorax must be penetrated. This process, called **thoracentisis** (thō-ra-sen-TĒ-sis) or **thoracocentesis** (thō-ra-kō-sen-TĒ-sis; *kentesis*, perforating), involves the penetration of the thoracic wall along the superior border of one of the ribs. Penetration at this location avoids damaging vessels and nerves within the costal groove. (See Figure 8-20b on p. 262 of the text.)

Spina Bifida

Spina bifida (SPĪ-na BI-fi-da) results when the vertebral laminae fail to unite during development. The neural arch is incomplete, and the membranes that line the dorsal body cavity bulge outward. In mild cases, most often involving the sacral and lumbar regions, the condition may pass unnoticed. In severe cases, the entire spinal column and skull are affected. This condition is often associated with developmental abnormalities of the brain and spinal cord. (See the Embryology Summary for Chapter 13 of the text, p. 418.)

CHAPTER 9
The Skeletal System: Appendicular Division

Hip Fractures and the Elderly
Page 286

Hip fractures are most often suffered by individuals over 60 years of age, when osteoporosis has weakened the thigh bones. These injuries may be accompanied by dislocation of the hip or pelvic fractures. For individuals with osteoporosis, healing proceeds very slowly. In addition, the powerful muscles that surround the joint can easily prevent proper alignment of the bone fragments. Trochanteric fractures usually heal well, if the joint can be stabilized; steel frames, pins, screws, or some combination of those devices may be needed to preserve alignment and permit healing to proceed normally.

Severe fractures of the femoral neck have the highest complication rate of any fracture because the blood supply to the region is relatively delicate. The surgical procedures used to treat trochanteric fractures are often unsuccessful in stabilizing femoral neck fractures. If more complex pinning operations fail, the entire joint can be replaced. In this "total hip" procedure, the damaged portion of the femur is removed. An artificial femoral head and neck is attached by a spike that extends into the marrow cavity of the shaft. Special cement is used to anchor it in place and to attach a new articular surface to the acetabulum. A photograph of an artificial hip was included in Figure 7-16 of the text (p. 229), in the discussion of artificial joints. (See also the Health News box on p. 288 of the text.)

Problems with the Ankle and Foot
Page 290

The ankle and foot are subjected to a variety of stresses during normal daily activities. In a **sprain**, a ligament is stretched to the point where some of the collagen fibers are torn. The ligament remains functional, and the structure of the joint is not affected. The most common cause of a sprained ankle is a forceful inversion of the foot that stretches the lateral ligament. An ice pack is usually required to reduce swelling, and with rest and support the ankle should heal in about 3 weeks.

In more serious incidents, the entire ligament may be torn apart, or the connection between the ligament and the malleolus may be so strong that the bone breaks before the ligament. In general, a broken bone heals more quickly and effectively than does a torn ligament. A dislocation often accompanies such injuries.

In a *dancer's fracture* the proximal portion of the fifth metatarsal is broken. This usually occurs while the body weight is being supported by the longitudinal arch, as when dancing on the toes. A sudden shift in weight from the medial portion of the arch to the lateral, less elastic border breaks the fifth metatarsal close to its distal articulation.

Individuals with abnormal arch development are more likely to suffer metatarsal injuries. Someone with **flat feet** loses or never develops the longitudinal arch. "Fallen arches" may develop as tendons and ligaments stretch and become less elastic. Obese individuals or those who must constantly stand or walk on the job are likely candidates. Children have very mobile articulations and elastic ligaments, so they often have **flexible flat feet**. Their feet look flat only while they are standing, and the arch appears when they stand on their toes or sit down. This condition usually disappears as growth continues.

Congenital **talipes equinus**, or *clubfoot*, results from an inherited developmental abnormality that affects 0.2 percent of births (2 in 1000). One or both feet may be affected, and the problem involves abnormal muscle development that distorts growing bones and joints. Usually the tibia, ankle, and foot are affected, and the feet are turned medially and inverted. The longitudinal arch is exaggerated, and if both feet are involved, the soles face one another. Prompt treatment with casts in infancy helps to alleviate the problem, and fewer than half of the cases require surgery; 1992 gold medal Olympic skater Kristi Yamaguchi was born with clubfeet.

Clawfeet are also produced by muscular abnormalities. In this case the median longitudinal arch becomes exaggerated because the plantar flexors are overpowering the dorsiflexors. Muscle cramps or nerve paralysis may be responsible; the condition tends to develop in adults, and it gets progressively worse with age.

CHAPTER 10

The Muscular System: Skeletal Muscle Tissue

Trichinosis

Page 298

Trichinosis (trik-i-NŌ-sis; *trichos*, hair + *nosos*, disease) results from infection by a parasitic nematode worm, *Trichinella spiralis*. Symptoms include diarrhea, weakness, and muscle pain. The muscular symptoms are caused by the invasion of skeletal muscle tissue by larval worms, which create small pockets within the permysium and endomysium. Muscles of the tongue, eyes, diaphragm, chest, and lower leg are most often affected.

Larvae are common in the flesh of pigs, horses, dogs, and other mammals. People are most often exposed by eating undercooked pork (the larvae are killed by cooking). Once eaten, the larvae mature within the intestinal tract, where they mate and produce eggs. The new generation of larvae then migrates through the body tissues to reach the muscles, where they complete their early development. The migration and subsequent settling produce a generalized achiness, muscle and joint pain, and swelling in infected tissues. An estimated 1.5 million Americans carry *Trichinella* around in their muscles, and up to 300,000 new infections occur each year. The mortality rate for people who have symptoms severe enough to require treatment is approximately 1 percent.

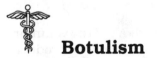

Botulism

Page 308

Botulinus (bot-ū-LĪ-nus) **toxin** prevents the release of ACh at the synaptic knob. It thus produces a severe and potentially fatal paralysis of skeletal muscles. A case of botulinus poisoning is called **botulism**.[1] The toxin is produced by a bacterium, *Clostridium botulinum*, that does not need oxygen to grow and reproduce. Because the organism can live quite happily in a sealed can or jar, most cases of

---

[1]This disorder was described 200 years ago by German physicians treating patients poisoned by dining on contaminated sausages. *Botulus* is the Latin word for sausage.

botulism are linked to improper canning or storing procedures, followed by failure to adequately cook the food before eating. Canned tuna or beets, smoked fish, and cold soups have most often been involved with cases of botulism. Boiling for 10 minutes destroys the toxin. Bacterial spores are more resistant, and may survive several hours of boiling.

Symptoms usually begin 12–36 hours after eating a contaminated meal. The initial symptoms are often disturbances in vision, such as seeing double or a painful sensitivity to bright lights. These symptoms are followed by other sensory and motor problems, including blurred speech and an inability to stand or walk. Roughly half of botulism patients experience intense nausea and vomiting. These symptoms persist for a variable period (days to weeks), followed by a gradual recovery; some patients are still recovering after a year.

The major risk of botulinus poisoning is respiratory paralysis and death by suffocation. Treatment is supportive: bed rest, observation, and, if necessary, use of a mechanical respirator. In severe cases drugs that promote the release of ACh, such as *guanidine hydrochloride*, may be administered. The overall mortality rate in the United States is about 10 percent.

 Myasthenia Gravis *Page 308*

Myasthenia gravis (mī-as-THĒ-nē-a GRA-vis) is characterized by a general muscular weakness that is often most pronounced in the muscles of the arms, head, and chest. The first symptom is usually a weakness of the eye muscles and drooping eyelids. Facial muscles are often weak as well, and the individual develops a peculiar smile known as the "myasthenic snarl." As the disease progresses, pharyngeal weakness leads to problems with chewing and swallowing, and it becomes difficult to hold the head upright.

The muscles of the upper chest and arms are next to be affected. All of the voluntary muscles of the body may ultimately be involved. Severe myasthenia gravis produces respiratory paralysis, with a mortality rate of 5–10 percent. However, the disease does not always progress to such a life-threatening stage. For example, roughly 20 percent of patients experience eye problems but no other symptoms.

The condition results from a decrease in the number of ACh receptors on the motor end plate. Before the remaining receptors can be stimulated enough to trigger a strong contraction, the ACh molecules are destroyed by cholinesterase. As a result, muscular weakness develops.

20

The primary cause of myasthenia gravis appears to be a malfunction in the immune system. The body attacks the ACh receptors of the motor end plate as if they were foreign proteins. For unknown reasons, women are affected twice as often as men. Estimates of the incidence of this disease in the United States range from 2 to 10 cases per 100,000 population.

One approach to therapy involves the administration of drugs, such as Neostigmine, that are termed **cholinesterase inhibitors**. These compounds, as their name implies, are enzyme inhibitors—they tie up the active sites at which cholinesterase normally binds ACh. With cholinesterase activity reduced, the concentration of ACh at the synapse can rise enough to stimulate the surviving receptors and produce muscle contraction.

Duchenne's Muscular Dystrophy *Page 310*

The muscular dystrophies (DIS-trō-fēz) are congenital diseases that produce progressive muscle weakness and deterioration. One of the most common and best understood conditions is **Duchenne's muscular dystrophy (DMD)**.

This form of muscular dystrophy appears in childhood, often between the ages of 3 and 7. The condition generally affects only males. A progressive muscular weakness develops, and the individual usually dies before age 20 because of respiratory paralysis. Skeletal muscles are primarily affected, although for some reason the facial muscles continue to function normally. In later stages of the disease the facial muscles and cardiac muscle tissue may also become involved.

The skeletal muscle fibers in a DMD patient are structurally different from those of normal individuals. Abnormal membrane permeability, cholesterol content, rates of protein synthesis, and enzyme composition have been reported. DMD sufferers also lack a protein, called *dystrophin*, found in normal muscle fibers. Although the functions of this protein remain uncertain, researchers have recently identified and cloned the gene for dystrophin. Rats with DMD have been cured by insertion of this gene into their muscle fibers—a technique that may eventually be used to treat human patients.

The inheritance of DMD is sex-linked: women carrying the defective genes are unaffected, but each of their male children will have a 50 percent chance of developing DMD. Now that the specific location of the gene has been identified, it is possible to determine whether a woman is carrying the defective gene or not. Chapter 29 of the text examines the genetic basis for DMD in greater detail (see pp. 958-959).

Polio

Because skeletal muscles depend on their motor neurons for stimulation, disorders that affect the nervous system can have an indirect affect on the muscular system. The **poliovirus** attacks and kills motor neurons in the spinal cord and brain. As the neurons die, the dependent motor units first become paralyzed and then undergo atrophy. The resulting condition is called **polio**. In severe cases, paralysis of the respiratory muscles makes breathing impossible without the assistance of an "iron lung" or some comparable device (Figure 10-A).

Polio has been almost completely eliminated from the U.S. population through a successful immunization program. In 1954 there were 18,000 new cases; there were 8 in 1976. Unfortunately, some parents refuse to immunize their children against the poliovirus on the assumption that the disease has been "conquered." This is a mistake because (1) there is no cure for polio, (2) the virus remains in the environment, and (3) approximately 38 percent of children ages 1–4 have not been fully immunized. Thus a major epidemic could still occur.

FIGURE 10-A
A polio victim in an iron lung

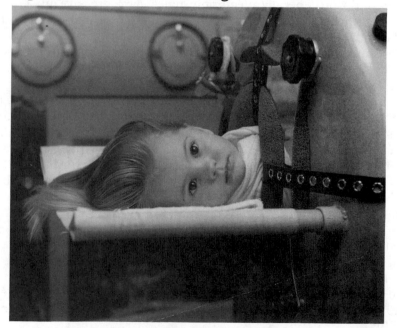

Fibromyalgia (*-algia*, pain) is an inflammatory disorder that has been recognized only in the last 10 years. Although first described in the early 1800s, the condition is still somewhat controversial because the reported symptoms cannot be linked to any anatomical or physiological abnormalities. However, physicians now recognize a distinctive pattern of symptoms that warrant consideration as a clinical entity.

Fibromyalgia may be the most common musculoskeletal disorder affecting women under 40 years of age. There may be 3–6 million cases in the United States today. Symptoms include chronic aches, pain, and stiffness, and multiple tender points at specific, characteristic locations. The four most common tender points are (1) just below the kneecap, (2,3) distal to the medial and lateral epicondyles of the humerus, and (4) the junction between the second rib and the cartilage attaching it to the sternum. An additional clinical criterion is that the pains and stiffness cannot be explained by other mechanisms. Patients suffering from this condition invariably report chronic fatigue; they feel tired on awakening and often complain of awakening repeatedly during the night.

Most of these symptoms could be attributed to other problems. For example, chronic depression can lead to fatigue and poor-quality sleep. As a result, the pattern of tender points is really the diagnostic key to fibromyalgia. This symptom clearly distinguishes fibromyalgia from **chronic fatigue syndrome (CFS).** As in fibromyalgia, individuals with CFS usually report fatigue, a general lack of energy, muscle aches, and difficulty sleeping. In addition, CFS may be associated with chronic swollen lymph nodes and low grade fevers. Attempts to link either fibromyalgia or CFS to viral infections or to some physical or psychological trauma have not been successful, and the cause remains a mystery. Treatment at present is limited to relieving symptoms when possible. For example, anti-inflammatory medications may help relieve pain, antidepressant drugs can be used to promote sleep, and exercise programs may help maintain normal fitness and range of motion.

CHAPTER 12
The Nervous System: Neural Tissue

Axoplasmic Transport and Disease   *Page 373*

With a soft flutter of wings, dark shapes drop from the sky onto the backs of grazing cattle. Each shape is a small bat whose scientific name, *Desmodus rotundus*, is less familiar than the popular term "vampire bat." Vampire bats inhabit tropical and semitropical areas of North, Central, and South America. They range from the Texas coast to Chile and southern Brazil. These rather aggressive animals are true vampires, subsisting on a diet of fresh blood. Over the next hour every bat in the flight—which may number in the hundreds—will consume about 65 mℓ of blood through small slashes in the skin of their prey.

As unpleasant as this blood collection may sound, it is not the blood loss that is the primary cause for concern. The major problem is that these bats can be carriers for the rabies virus. **Rabies** is an acute disease of the central nervous system. The rabies virus can infect any mammal, wild or domestic, and with few exceptions the result is death within 3 weeks. (For unknown reasons, bats can survive rabies infection for an indefinite period. As a result, an infected bat can serve as a carrier for the disease. Because many bat species, including vampires, form dense colonies, a single infected individual can spread the disease through the entire colony.)

The deaths of cattle and pigs are economic hardships; human infection is a disaster. Rabies is usually transmitted to people through the bite of a rabid animal. There are an estimated 15,000 cases of rabies each year worldwide, the majority of them the result of dog bites. Only about five of those cases, however, are diagnosed in the United States, and since dogs and cats in the United States are vaccinated against rabies, cases here are most commonly caused by the bites of raccoons, foxes, skunks, and bats.

Although these bites usually involve peripheral sites, such as the hand or foot, the symptoms are caused by CNS damage. The virus present at the injury site is absorbed by synaptic knobs in the region. It then gets a free ride to the CNS, courtesy of retrograde flow. Over the first few days following exposure, the individual

24

experiences headache, fever, muscle pain, nausea, and vomiting. The victim then enters a phase marked by extreme excitability, with hallucinations, muscle spasms, and disorientation. There is difficulty in swallowing, and the accumulation of saliva makes the individual appear to be "foaming at the mouth." Coma and death soon follow.

Preventive treatment, which must begin almost immediately following exposure, consists of injections containing antibodies against the rabies virus. This postexposure treatment may not be sufficient following massive infection, which can lead to death in as little as 4 days. Individuals such as veterinarians or field biologists who are at high risk for exposure often take a preexposure series of injections. These injections bolster the immune defenses and improve the effectiveness of the postexposure treatment. *Without treatment, rabies infection is always fatal.*

Rabies is perhaps the most dramatic example of a clinical condition directly related to axoplasmic flow. However, many toxins, including heavy metals, some pathogenic bacteria, and other viruses use this mechanism to enter the CNS.

Neurotoxins in Seafood *Page 385*

Several forms of human poisoning result from eating seafood containing *neurotoxins*, poisons that primarily affect neurons. **Tetrodotoxin** (te-TRŌ-dō-tok-sin), or **TTX**, is found in the liver, gonads, and blood of certain Pacific puffer fish species, and a related compound is found in the skin glands of some salamanders. Tetrodotoxin selectively blocks voltage-regulated sodium ion channels, effectively preventing nerve cell activity. The result is usually death from paralysis of the respiratory muscles. (Despite the risks, the Japanese consider the puffer fish a delicacy, served under the name *fugu*. Specially licensed chefs prepare these meals, carefully removing the potentially toxic organs. Nevertheless, a mild tingling and sense of intoxication is considered desirable, and several people die each year as a result of improper preparation of the dish.)

Saxitoxin (sak-si-TOK-sin), or **STX**, can have a similarly lethal effect. Saxitoxin and related poisons are produced by several species of marine microorganisms. When these organisms undergo a population explosion they color the surface waters, producing a *red tide*. Eating seafood that has become contaminated by feeding on the toxic microbes can result in the characteristic symptoms of **paralytic shellfish poisoning** (**PSP**; from clams, mussels, or oysters) or **ciguatera** (**CTX**; from fish).

Growth and Myelination of the Nervous System

Page 386

The development of the nervous system continues for many years after birth. Nerve cells increase in number for the first year after delivery, and most of the important interconnections between neurons occur after birth, rather than before. Growth of the brain is completed by age 4, but the neurons have yet to be interconnected extensively. The myelination of axons may not be completed until early adolescence. The level of nervous system development limits mental and physical performance. For example, the degree of interconnection between neurons in the CNS affects intellectual abilities, and the myelination of axons improves coordination and control by decreasing the time between reception of a sensation and completion of a response.

Demyelination Disorders

Page 386

Demyelination is the progressive destruction of myelin sheaths in the CNS and PNS. The result is a gradual loss of sensation and motor control that leaves affected regions numb and paralyzed. Many unrelated conditions can cause demyelination; four important examples of demyelinating disorders are *heavy metal poisoning*, *diphtheria*, *multiple sclerosis*, and *Guillain-Barré syndrome*.

Heavy Metal Poisoning Chronic exposure to heavy metal ions, such as arsenic, lead, or mercury, can lead to glial cell damage and demyelination. As demyelination occurs, the affected axons deteriorate, and the condition becomes irreversible. Historians note several interesting examples of heavy metal poisoning with widespread impact. For example, lead contamination of drinking water has been cited as one factor in the decline of the Roman Empire. In the seventeenth century, the great physicist Sir Isaac Newton is thought to have suffered several episodes of physical illness and mental instability brought on by his use of mercury in chemical experiments. Well into the nineteenth century, mercury used in the preparation of felt presented a serious occupational hazard for those employed in the manufacture of stylish hats. Over time, mercury absorbed through the skin and across the lungs accumulated in the CNS, producing neurological damage that affected both physical and mental function. (This effect is the source of the expression "mad as a hatter.") More recently, Japanese fishermen working in Minamata Bay, Japan, collected and consumed seafood contaminated with mercury discharged from a nearby chemical plant. Levels of mercury in their systems gradually rose to the point that clinical symptoms appeared in hundreds of people. Making matters worse, mercury contamination of developing embryos caused severe, crippling birth defects.

Diphtheria   Diphtheria (dif-THĒ-rē-a; *diphthera*, membrane + *-ia*, disease) is a disease that results from a bacterial infection of the respiratory tract. In addition to restricting airflow and sometimes damaging the respiratory surfaces, the bacteria produce a powerful toxin that injures the kidneys and adrenal glands, among other tissues. In the nervous system, diphtheria toxin damages Schwann cells and destroys myelin sheaths in the PNS. This *demyelination* leads to sensory and motor problems that may ultimately produce a fatal paralysis. The toxin also affects cardiac muscle cells, and heart enlargement and failure may occur. The fatality rate for untreated cases ranges from 35 to 90 percent, depending on the site of infection and the subspecies of bacterium. Because an effective vaccine exists, cases are relatively rare in countries with adequate health care.

Multiple Sclerosis   Multiple sclerosis (skler-Ō-sis; *sklerosis*, hardness), or **MS**, is a disease characterized by recurrent incidents of demyelination affecting axons in the optic nerve, brain, and/or spinal cord. Common symptoms include partial loss of vision and problems with speech, balance, and general motor coordination. The time between incidents and the degree of recovery vary from case to case. In about one-third of all cases the disorder is progressive, and each incident leaves a greater degree of functional impairment. The average age at the first attack is 30–40; the incidence in women is 1.5 times that among men. There is no effective treatment at present. MS is discussed further in Chapter 13 of this Manual.

Guillain-Barré Syndrome   Guillain-Barré syndrome is characterized by a progressive demyelination of somatic motor neurons. Symptoms initially involve weakness of the legs, which spreads rapidly to muscles of the trunk and arms. These symptoms usually increase in intensity for 1–2 weeks before subsiding. The mortality rate is low (under 5 percent), but there may be some permanent loss of motor function. The cause is unknown, but because roughly two-thirds of Guillain-Barré patients develop symptoms within 2 months after a viral infection, it is suspected that the condition may result from a malfunction of the immune system. (The mechanism involved is considered in Chapter 22 of the text—see p. 741.)

Tay-Sachs Disease *Page 398*

Tay-Sachs disease is a genetic abnormality involving the metabolism of gangliosides, important components of nerve cell membranes. It is a *lysosomal storage disease* (see the discussion in Chapter 3 of this Manual)—its victims lack an enzyme needed to break down one particular ganglioside, which accumulates within the lysosomes of CNS neurons and causes them to deteriorate. Affected infants seem normal at birth, but within 6 months neurological problems begin to appear. The progress of symptoms typically includes muscular weakness, blindness, seizures, and death, usually before age 4. No effective treatment exists, but prospective parents can be tested to determine whether or not they are carrying the gene responsible for this condition. The disorder is most prevalent in one ethnic group, the Ashkenazi Jews of Eastern Europe.

CHAPTER 13

The Nervous System: The Spinal Cord and Spinal Nerves

Shingles
Page 406

In **shingles**, or **Herpes zoster**, the Herpes varicella-zoster virus attacks neurons within ganglia of the dorsal roots and cranial nerves. This disorder produces a painful rash whose distribution corresponds to that of the affected sensory nerves. Shingles develops in adults who were first exposed to the virus as children. The initial infection produces symptoms known as *chickenpox*. After this encounter the virus remains dormant within neurons of the anterior gray horns. It is not known what triggers reactivation of this pathogen. Fortunately for those affected, attacks of shingles usually heal and leave behind only unpleasant memories.

Most people suffer only a single episode of shingles in their adult lives. However, the problem may recur in people with weakened immune systems, including those with AIDS or some forms of cancer. Treatment typically involves large doses of the antiviral drug *acyclovir (Zovirax)*.

Spinal Meningitis
Page 406

Meningitis is the inflammation of the meningeal membranes. Meningitis often disrupts the normal circulatory and cerebrospinal fluid supplies, damaging or killing neurons and glial cells in the affected areas. Although the initial diagnosis may specify the meninges of the spinal cord (**spinal meningitis**) or brain (**cerebral meningitis**), in later stages the entire meningeal system is usually affected.

The warm, dark, nutrient-rich environment of the meninges provides ideal conditions for a variety of bacteria and viruses. Microorganisms that cause meningitis include those associated with middle ear infections, pneumonia, streptococcal ("strep"), staphylococcal ("staph"), or meningococcal infections, and tuberculosis. These pathogens may gain access to the meninges by traveling within blood vessels or by entering at sites of vertebral or cranial injury. Headache, chills, high fever, disorientation, and rapid heart and respiratory rates appear as higher centers are affected. Without treatment, delirium, coma, convulsions, and death may follow within hours.

The most common clinical assessment involves checking for a "stiff neck" by asking the patient to touch chin to chest. Meningitis affecting the cervical portion of the spinal cord results in an increase in the muscle tone of the extensor muscles of the neck. So many motor units become activated that voluntary or involuntary flexion of the neck becomes painfully difficult if not impossible.

The mortality rate for viral meningitis ranges from 1 to 50 percent or higher, depending on the type of virus, the age and health of the patient, and other factors. There is no effective treatment for viral meningitis, but bacterial meningitis can be combatted with antibiotics and the maintenance of proper fluid and electrolyte balance.

Spinal Anesthesia   *Page 408*

Local anesthetics introduced into the subarachnoid space of the spinal cord produce a temporary sensory and motor paralysis. The effects spread as the anesthetic diffuses along the cord, and precise control of the regional effects can be difficult to achieve. Problems with overdosing are seldom serious, because the diaphragmatic breathing muscles are controlled by upper cervical spinal nerves. Thus respiration continues even when the thoracic and abdominal segments have been paralyzed.

Sometimes more precise control can be obtained by injecting the anesthetic into the epidural space, producing an **epidural block**. This technique has the advantage of affecting only the spinal nerves in the immediate area of the injection. Epidural anesthesia may be difficult to achieve, however, in the upper cervical, midthoracic, and lumbar regions, where the epidural space is extremely narrow. **Caudal anesthesia** involves the introduction of anesthetics into the epidural space of the sacrum. Injection at this site paralyzes lower abdominal and perineal structures. Epidural blocks in the lower lumbar or sacral regions or caudal anesthesia may be used to control pain during childbirth.

Multiple Sclerosis   *Page 412*

Multiple sclerosis (MS), introduced in Chapter 12 of this Manual (see Demyelination Disorders), is a disease that produces muscular paralysis and sensory losses through demyelination. The initial symptoms appear as the result of myelin degeneration within the white matter of the lateral and posterior columns of the spinal cord or along tracts within the brain. For example, spinal cord involvement may produce weakness, tingling sensations, and a loss of "position sense" for the arms or legs. During subsequent attacks the effects become more widespread, and the cumulative sensory and motor losses may eventually lead to a generalized muscular paralysis.

Recent evidence suggests that this condition may be linked to a defect in the immune system that causes it to attack myelin sheaths. MS patients have lymphocytes that do not respond normally to foreign proteins, and because several viral proteins have amino acid sequences similar to those of normal myelin, it has been proposed that MS results from a case of mistaken identity. For unknown reasons MS appears to be associated with cold and temperate climates. It has been suggested that individuals developing MS may have an inherited susceptibility to the virus that is exaggerated by environmental conditions. The yearly incidence within the United States averages around 50 cases for every 100,000 in the population. There is as yet no effective treatment.

Palsies   *Page 417*

Peripheral nerve **palsies,** also known as **peripheral neuropathies**, are characterized by regional losses of sensory and motor function as the result of nerve trauma or compression. **Brachial palsies** result from injuries to the brachial plexus or its branches. **Crural palsies** involve the nerves of the lumbosacral plexus.

Although palsies may appear for several reasons, the **pressure palsies** are especially interesting. A familiar but mild example is the experience of having an arm or leg "fall asleep." The limb becomes numb, and afterwards an uncomfortable "pins-and-needles" sensation, or **paresthesia**, accompanies the return to normal function.

These incidents are seldom of clinical significance, but they provide graphic examples of the effects of more serious palsies that can last for days to months. In **radial nerve palsy**, pressure on the back of the arm interrupts the function of the radial nerve, so that the extensors of the wrist and fingers are paralyzed. This condition is also known as "Saturday night palsy," for falling asleep on a couch with your arm over the seat back (or beneath someone's head) can produce the right combination of pressures. Students may also be familiar with **ulnar palsy**, which can result from prolonged contact between elbow and desk. The ring and little fingers lose sensation, and the fingers cannot be adducted.

Men with large wallets in their hip pockets may develop symptoms of **sciatic compression** after driving or sitting in one position for extended periods. As nerve function declines, the individuals notice some lumbar pain, a numbness along the back of the leg, and a weakness in the leg muscles. Similar symptoms result from compression of the lumbar nerve roots by a distorted lumbar intervertebral disc. This condition is termed **sciatica**, and one or both legs may be affected, depending on the site of compression. Finally, sitting with your legs crossed may produce symptoms of a **peroneal palsy**. Sensory losses from the top of the foot and side of the lower leg are accompanied by a decreased ability to dorsiflex or evert the foot.

Leprosy   *Page 417*

The condition traditionally called **leprosy**, now more commonly known as **Hansen's disease**, is an infectious disease caused by a bacterium, *Mycobacterium leprae*. It is a disease that progresses slowly, and symptoms may not appear for up to 30 years after infection. The bacterium invades peripheral nerves, especially those in the skin, producing initial sensory losses. Over time motor paralysis develops, and the combination of sensory and motor loss can lead to recurring injuries and infections. The eyes, nose, hands, and feet may develop deformities as a result of neglected injuries. There are several forms of this disease; peripheral nerves are always affected, but some forms also involve extensive skin and mucous membrane lesions.

Only about 5 percent of those exposed develop symptoms; people living in the tropics are at greatest risk. There are about 2000 cases in the United States, and an estimated 12–20 million cases worldwide. The disease can usually be treated successfully with drugs such as *rifampin* and *dapsone*. Treated individuals are not infectious, and the practice of confining "lepers" in isolated compounds has been discontinued.

In **hyporeflexia** normal reflexes are weak, but apparent, especially with reinforcement. In **areflexia** (ā-re-FLEK-sē-a; *a-*, without) normal reflexes fail to appear, even with reinforcement. Hyporeflexia or areflexia may indicate temporary or permanent damage to skeletal muscles, dorsal or ventral nerve roots, spinal nerves, the spinal cord, or the brain.

Hyperreflexia occurs when higher centers maintain a high degree of facilitation along the spinal cord. Under these conditions reflexes are easily triggered, and the responses may be grossly exaggerated. This effect can also result from spinal cord compression or diseases that target higher centers or descending tracts. One potential result of hyperreflexia is the appearance of alternating contractions in opposing muscles. When one muscle contracts, it stimulates the stretch receptors in the other. The stretch reflex then triggers a contraction in that muscle, and this stretches receptors in the original muscle. This self-perpetuating sequence can be repeated indefinitely; it is called **clonus** (KLŌ-nus).

A more extreme hyperreflexia develops if the motor neurons of the spinal cord lose contact with higher centers. Often, following a severe spinal injury, the individual first experiences a temporary period of areflexia known as *spinal shock*, discussed in the Clinical Comment box on p. 411 of the text. When the reflexes return, they respond in an exaggerated fashion, even to mild stimuli. For example, the lightest touch on the skin surface may produce a massive withdrawal reflex. The reflex contractions may occur in a series of intense muscle spasms potentially strong enough to break bones. In the **mass reflex** the entire spinal cord becomes hyperactive for several minutes, issuing exaggerated skeletal muscle and visceral motor commands.

CHAPTER 14

The Nervous System: The Brain and Cranial Nerves

Cerebellar Dysfunction   *Page 449*

Cerebellar function may be permanently altered by trauma or a stroke, or temporarily by drugs such as alcohol. Such alterations can produce severe disturbances in motor control. **Ataxia** ("lack of order") refers to the disturbance of balance that in severe cases leaves the individual unable to stand without assistance. Less severe conditions cause an obvious unsteadiness and irregular patterns of movement. The individual often watches his or her feet to see where they are going and controls ongoing movements by intense concentration and voluntary effort. Reaching for something becomes a major exertion, for the only information available must be gathered by sight or touch while the movement is taking place. Without the cerebellar ability to adjust movements while they are occurring, the individual becomes unable to anticipate the time course of a movement. Most often, a reaching movement ends with the hand overshooting the target. This inability to anticipate and stop a movement precisely is called **dysmetria** (dis-MET-rē-a; *dys-*, bad + *metron*, measure). In attempting to correct the situation, the hand usually overshoots again in the opposite direction, and then again. This leaves the hand oscillating back and forth until either the object can be grasped or the attempt is abandoned. This oscillatory movement is known as an **intention tremor**.

Clinicians check for ataxia by watching an individual walk in a straight line; the usual test for dysmetria involves touching the tip of the index finger to the tip of the nose. Because many drugs impair cerebellar performance, the same tests are used by police officers to check drivers suspected of alcohol or other drug abuse.

Cranial Trauma

Page 451

Cranial trauma is a head injury resulting from harsh contact with another object. Head injuries account for over half of the deaths attributed to trauma each year. There are roughly 8 million cases of cranial trauma annually, and over a million of these are major incidents involving intracranial hemorrhaging, concussion, contusion, or laceration of the brain. The characteristics of spinal concussion, contusion, and laceration are introduced in the Clinical Comment box on p. 411 of the text, and the same descriptions can be applied to injuries of the brain.

Concussions usually accompany even minor head injuries. A concussion involves a temporary loss of consciousness and some degree of amnesia. Physicians examine any concussed individual quite closely and may X-ray the skull to check for skull fractures or cranial bleeding. Mild concussions produce a brief interruption of consciousness and little memory loss. Severe concussions produce extended periods of unconsciousness and abnormal neurological functions. Severe concussions are often associated with contusions (bruises) or lacerations (tears); the possibilities for recovery vary depending on the areas affected. Extensive damage to the reticular formation may produce a permanent state of unconsciousness, while damage to the lower brain stem will usually prove fatal.

Tic Douloureux

Page 463

Tic douloureux (doo-loo-ROO; painful) affects one individual out of every 25,000. Sufferers complain of severe, almost totally debilitating pain triggered by contact with the lip, tongue, or gums. The pain arrives with a sudden, shocking intensity and then disappears. Usually only one side of the face is involved. Another name for this condition is **trigeminal neuralgia**, for it is cranial nerve V that innervates the sensitive areas. This condition usually affects adults over 40 years of age; the cause is unknown. The pain can often be temporarily controlled by drug therapy, but surgical procedures may eventually be required. The goal of the surgery is the destruction of the afferents carrying the pain sensations. This can be attempted by actually cutting the nerve, a procedure called a **rhizotomy** (*rhiza*, root), or by injecting chemicals such as alcohol or phenol into the nerve at the foramina ovale and rotundum. The sensory fibers may also be destroyed by inserting an electrode and cauterizing the sensory nerve trunks as they leave the semilunar ganglion.

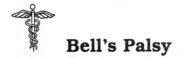

Bell's Palsy

Bell's palsy results from an inflammation of the facial nerve that is probably related to viral infection. Involvement of the facial nerve (VII) can be deduced from the symptoms: paralysis of facial muscles on the affected side (see Figure 14-A) and loss of taste sensations from the anterior two-thirds of the tongue. The individual does not show prominent sensory deficits, and the condition is usually painless. Most of the time Bell's palsy cures itself after a few weeks or months.

**FIGURE 14-A
Bell's palsy**

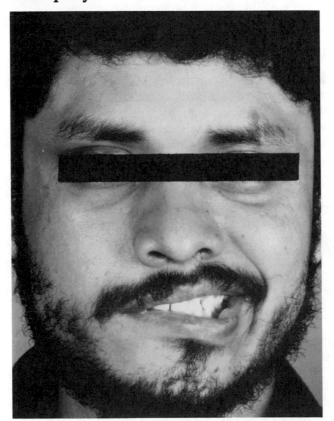

CHAPTER 15

The Nervous System: Pathways, Processing, and Higher-Order Functions

Cerebral Palsy

Page 477

The term **cerebral palsy** refers to a number of disorders affecting voluntary motor performance that appear during infancy or childhood and persist throughout the life of the affected individual. The cause may be trauma associated with premature or unusually stressful birth, maternal exposure to drugs, including alcohol, or a genetic defect that causes improper development of the motor pathways. Prematurity has become less of a factor as maintenance and support procedures have improved, but difficult labor and deliveries still pose a problem. If the oxygen concentration in the fetal blood declines significantly for as little as 5–10 minutes, CNS function may be permanently impaired. The cerebral cortex, cerebellum, cerebral nuclei, hippocampus, and thalamus are likely targets, producing abnormalities in motor performance, involuntary control of posture and balance, memory, speech, and learning abilities.

Seizures and Epilepsy

Page 484

A **seizure** is a temporary disorder of cerebral function, accompanied by abnormal, involuntary movements, unusual sensations, and/or inappropriate behavior. The individual may or may not lose consciousness for the duration of the attack. There are many different types of seizures. Clinical conditions characterized by seizures are known as *seizure disorders*, or *epilepsies*.

Seizures of all kinds are accompanied by a marked change in the pattern of electrical activity monitored in an electroencephalogram. The alteration begins in one portion of the cerebral cortex but may subsequently spread to adjacent regions, potentially involving the entire cortical surface. The neurons at the site of origin are abnormally sensitive. When they become active, they may facilitate and subsequently stimulate adjacent neurons. As a result, the abnormal electrical activity can spread across the cortex like a wave on the surface of a pond.

The extent of the cortical involvement determines the nature of the observed symptoms. A **focal seizure** affects a relatively restricted cortical area, producing either sensory or motor symptoms. The individual usually remains conscious throughout the attack. If the seizure occurs within a portion of the primary motor cortex, the activation of pyramidal cells will produce uncontrollable movements. The muscles affected or the specific sensations experienced provide an indication of the precise region involved. In a **temporal lobe seizure** the disturbance spreads to the sensory cortex and association areas, so the individual also experiences unusual memories, sights, smells, or sounds. Involvement of the limbic system may also produce sudden emotional changes. Often the individual will lose consciousness at some point during the incident.

Epilepsy, or **seizure disorders**, refers to more than 40 different conditions characterized by a recurring pattern of seizures over extended periods. In roughly 75 percent of patients, no obvious cause can be determined.

Convulsive seizures are associated with uncontrolled muscle contractions. In a *generalized seizure* the entire cortical surface is involved. Generalized seizures may range from prolonged, major events to brief, almost unnoticed incidents. Only two examples will be considered here, *grand mal* and *petit mal* seizures.

Most readers will think of an epileptic attack as involving powerful, uncoordinated muscular contractions affecting the face, eyes, and limbs. These are symptoms of a **grand mal** seizure. During a grand mal attack the cortical activation begins at a single focus and then spreads across the entire surface. There may be no warning, but some individuals experience a vague apprehension or awareness that a seizure is about to begin. There follows a sudden loss of consciousness, and the individual drops to the floor as major muscle groups go into tonic contraction. The body remains rigid for several seconds before a rhythmic series of contractions occurs in the arm and leg muscles. After the attack subsides, the individual may appear disoriented or sleep for several hours. Muscles or bones subjected to extreme stresses may be damaged, and the person will probably be rather sore for days after the incident.

Petit mal epileptic attacks are very brief (under 10 seconds in duration) and involve few motor abnormalities. Typically the individual simply loses consciousness suddenly, with no warning. It is as if an internal switch were thrown and the conscious mind turned off. Because the individual is "not there" for brief periods during petit mal attacks, the incidents are known as *absence seizures*. During the seizure there may be small motor activities, such as fluttering of the eyelids or trembling of the hands.

Petit mal attacks usually begin between ages 6 and 14. They can occur hundreds of times per day, so that the child lives each day in small segments separated by blank periods. The victim is aware of brief losses of consciousness that occur without warning, but seldom seeks help because of embarassment. Often he or she becomes extremely anxious about the timing of future attacks. However, the motor signs are so minor as to go completely unnoticed by other family members, and the psychological stress caused by this condition is often overlooked. The initial diagnosis is frequently made during counseling for learning problems. (You have probably taken an exam after missing one or two lectures of 20. Imagine taking an exam after missing every third minute of every lecture.)

Both petit mal and grand mal epilepsy can be treated with barbiturates or other anticonvulsive drugs, such as phenytoin sodium (*dilantin*).

PET Scans of the Brain   *Page 484*

Positron emission tomography (PET) scans, introduced in Chapter 2 of the text, can be used to monitor dynamic processes, such as changes in blood flow or metabolism over time, in specific parts of the brain. When a particular region of the brain becomes active, blood flow to that area increases, and so does the rate of glucose consumption. Thus, monitoring either of these characteristics in experimental subjects can provide information about the brain areas involved in specific

FIGURE 15-A
PET scans showing brain activity prompted by various auditory stimuli

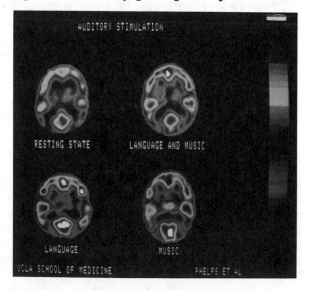

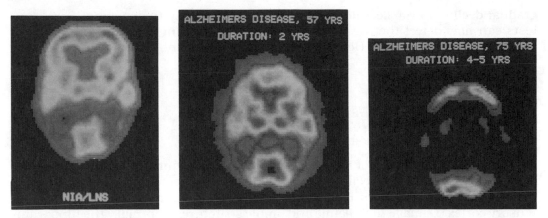

FIGURE 15-B
PET scans showing the effects of Alzheimer's disease on brain activity

functions. In a PET scanning procedure, the patient receives a small quantity of glucose labeled with a radioactive isotope, such as ^{11}O. When the blood flow to a particular region of the brain increases, so does the rate of glucose uptake and utilization. The radiation at that site increases markedly, and the PET scanner records the fact. Figure 15-A is a series of images showing how metabolic activity in different parts of the brain increases under the influence of various kinds of auditory stimuli.

PET scans can also register the changes in brain function that accompany various pathological states. In Figure 15-B, a normal PET scan is contrasted with scans of two individuals suffering from Alzeheimer's disease. The progressive deterioration in activity in several regions can be clearly seen.

Huntington's Disease   *Page 486*

Huntington's disease is an inherited disease marked by a progressive deterioration of mental abilities. There are racial differences in the incidence of this condition; Caucasians have by far the highest incidence, 3–4 cases per million population.

In Huntington's disease the cerebral nuclei show degenerative changes, as do the frontal lobes of the cerebral cortex. The basic problem is the destruction of ACh-secreting and GABA-secreting neurons in the cerebral nuclei. The cause of this deterioration is not known. The first signs of the disease usually appear in early adulthood. As you would expect in view of the areas affected, the symptoms involve difficulties in performing voluntary and involuntary patterns of movement and a gradual decline in intellectual abilities leading eventually to dementia. Screening tests can now detect the presence of the gene for Huntington's disease, which is located on chromosome 4. However, no effective treatment is available.

Amnesia *Page 490*

Amnesia refers to the loss of memory from disease or trauma. The type of memory loss depends upon the specific regions of the brain affected. Damage to sensory association areas produces memory loss of sensations arriving at the adjacent sensory cortex. Damage to thalamic and limbic structures, especially the hippocampus, will affect memory storage and consolidation. Amnesia may occur suddenly or progressively, and recovery may be complete, partial, or nonexistent, depending on the nature of the problem.

In **retrograde amnesia** (*retro-*, behind) the individual loses memories of past events. Some degree of retrograde amnesia often follows a head injury, and accident victims are frequently unable to remember the moments preceding a car wreck. In **anterograde amnesia** (*antero-*, ahead) an individual may be unable to store additional memories, but earlier memories are intact and accessible. The problem appears to involve an inability to generate long-term memories. At least two drugs—*Valium*® (*diazepam*) and *Halcion*®—have been known to cause brief periods of anterograde amnesia. A person with permanent anterograde amnesia lives in surroundings that are always new. Magazines can be read, chuckled over, and then reread a few minutes later with equal pleasure, as if they had never been seen before. Physicians and nurses must introduce themselves at every meeting, even if they have been visiting the patient for years.

Post-traumatic amnesia (PTA) often develops after a head injury. The duration of the amnesia varies depending on the severity of the injury. PTA combines the characteristics of retrograde and anterograde amnesia; the individual can neither remember the past nor consolidate memories of the present.

Sleep Disorders *Page 492*

Sleep disorders include abnormal patterns of REM or deep sleep, variations in the time of onset or the time devoted to sleeping, and unusual behaviors performed while sleeping. Sleep disorders of one kind or another are very common, affecting an estimated 25 percent of the United States population.

Many clinical conditions will affect sleep patterns or be exaggerated by the autonomic changes that accompany the various stages of sleep. Major clinical categories of sleep disorders include *parasomnias*, *insomnias*, and *hypersomnias*.

Parasomnias are abnormal behaviors performed during sleep. Sleepwalking, sleep talking, teeth-grinding, and so forth often involve slow wave sleep, when the skeletal muscles are not maximally inhibited. Parasomnias may have psychological rather than physiological origins.

Insomnia is characterized by shortened sleeping periods and difficulty in getting to sleep. This is the most common sleep disorder—an estimated one-third of adults suffer from insomnia. Temporary insomnia may accompany stress, such as family arguments or other crises. Insomnia of longer duration can result from chronic depression, illness, or drug abuse. Treatment varies, depending on the primary cause of the insomnia. For example, treatment of mild, temporary insomnia may involve an exercise program and reduction of caffeine intake. Severe insomnia can be treated with drugs, usually *benzodiazepines* such as *temazepam*, *flurazepam*, or *triazolam*.

Hypersomnia involves extremely long periods of otherwise normal sleep. The individual may sleep until noon and nap before dinner, despite an early retirement in the evening. These conditions may have physiological or psychological origins, and successful treatment may involve drug therapy or counseling or both. Two important examples of hypersomnias are *narcolepsy* and *sleep apnea*.

Roughly 0.2–0.3 percent of the population suffers from **narcolepsy**, characterized by dropping off to sleep at inappropriate times. The sleep lasts only a few minutes and is preceded by a period of muscular weakness. Any exciting stimulus, such as laughter or other strong emotion, may trigger the attack. (Interestingly, this condition occurs in other mammals; some dogs will keel over in a sound sleep when presented with their favorite treats.) In some instances, the sleep period is followed by a brief period of amnesia. Drugs that block REM sleep, such as methylphenidate (*Ritalin*®), will often reduce or eliminate narcoleptic attacks.

In **sleep apnea** a sleeping individual repeatedly stops breathing for short periods of time, probably due to a disturbance or abnormality in the respiratory control mechanism. The problem can be exaggerated by various drugs (including alcohol), obesity, hypertension, tonsilitis, and other medical conditions that affect either the CNS or the respiratory passageways. Roughly half of those experiencing sleep apnea show simultaneous cardiac arrhythmias, but the link between the two is not understood. Treatment focuses primarily on alleviating potential causes—for example, overweight individuals are put on a diet, and inflamed tonsils are removed.

CHAPTER 16

The Nervous System: Autonomic Division

Autonomic Tone and Hypersensitivity *Page 515*

In **Horner's syndrome** the sympathetic postganglionic innervation to one side of the face becomes interrupted. This may occur as the result of an injury, a tumor, or some progressive condition such as multiple sclerosis. The affected side of the face becomes flushed because in the absence of sympathetic tone the blood vessels dilate. Sweating stops in the affected region, and the pupil on that side becomes markedly constricted. Other symptoms include a drooping eyelid and an apparent retreat of the eye into the orbit.

The elimination of sympathetic innervation can have additional consequences that appear more gradually. Under normal conditions, sympathetic tone provides the effectors with a background level of stimulation. Following the disappearance of sympathetic stimulation, the effectors may become extremely sensitive to norepinephrine and epinephrine. This hypersensitivity can produce changes in facial blood flow and other functions when the adrenal medulla is stimulated.

CHAPTER 17

Sensory Function

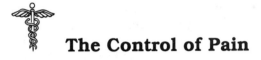 **The Control of Pain** *Page 528*

Pain management poses a number of problems for clinicians. Painful sensations can result from tissue damage or sensory nerve irritation; it may originate where it is perceived, be referred from another location, or represent a "false" signal generated along the sensory pathway. The treatment differs in each case, and an accurate diagnosis is an essential first step.

When pain results from tissue damage, the most effective solution is to stop the damage and end the stimulation. This is not always possible. Alternatively, the painful sensations can be suppressed at the injury site. Topical or locally injected anesthetics inactivate nociceptors in the immediate area. Aspirin and related analgesics reduce inflammation and suppress the release of irritating chemicals, such as enzymes or prostaglandins, in damaged tissues.

Pain can also be suppressed by inhibition of the pain pathway. Analgesics related to morphine reduce pain by mimicking the action of endorphins. Surgical steps can be taken to control severe pain, including: (1) the sensory innervation of an area can be destroyed by an electric current; (2) the dorsal roots carrying the painful sensations can be cut (a **rhizotomy**); (3) the ascending tracts can be severed (a **tractotomy**); or (4) thalamic or limbic centers can be stimulated or destroyed. These options, listed in order of increasing degree of effect, surgical complexity, and associated risk, are used only when other methods of pain control have failed to provide relief.

Many aspects of pain generation and control remain a mystery. Some patients experience a significant reduction in pain after receiving a nonfunctional medication. It has been suggested that this "placebo effect" results from endorphin release triggered by the expectation of pain relief. The Chinese technique of **acupuncture** to control pain has recently received considerable attention. Fine needles are inserted at specific locations and are either heated or twirled by the therapist. Several theories have been proposed to account for the positive effects, but none is widely accepted. The acupuncture points do not correspond to the distribution of any of the major peripheral nerves, and pain relief may come from the central release of endorphins.

Otitis Media and Mastoiditis Page 538

Otitis media is an infection of the middle ear, most often of bacterial origin. **Acute otitis media** typically affects infants and children—it is less often seen in adults. The pathogens gain access by ascending the Eustachian tube, usually after invading the nasal lining and producing an upper respiratory infection. As the pathogen population rises, white blood cells rush to the site, and the middle ear becomes filled with pus. Eventually the tympanic membrane may rupture, producing a characteristic oozing from the external auditory meatus. The bacteria can usually be controlled by antibiotics, the pain reduced by analgesics, and the swelling reduced by decongestants. In the United States it is rare for otitis media to progress to the stage at which tympanic membrane rupture occurs.

Otitis media is extremely common in underdeveloped countries where medical care and antibiotics are not readily available. Both children and adults in these countries often suffer from **chronic otitis media**, characterized by recurring bouts of infection. This condition produces scarring at the tympanic membrane, which often leads to some degree of hearing loss. Damage to the inner ear or the auditory ossicles may also reduce auditory sensitivity.

If the pathogens leave the middle ear and invade the air cells within the mastoid process, **mastoiditis** develops. The connecting passageways are very narrow, and as the infection progresses the subject experiences severe earaches, fever, and swelling behind the ear in addition to the symptoms of otitis media. The same antibiotic treatment is used to deal with both conditions, the particular antibiotic selected depending on the identity of the bacterium involved. The major risk of mastoiditis is the spread of the infection, by way of the connective tissue sheath of the facial nerve (N VII), into the cranial cavity and brain. Prompt antibiotic therapy is needed, and if the problem remains, the patient may have to undergo mastoidectomy (opening and drainage of the mastoid sinuses) or myringotomy (drainage of the middle ear through a surgical opening in the tympanic membrane).

Vertigo, Dizziness, and Motion Sickness Page 540

The term **vertigo** describes an inappropriate sense of motion. This distinguishes it from "dizziness," a sensation of lightheadedness and disorientation that often precedes a fainting spell. Vertigo can result from disturbances in central processing or abnormal conditions at the peripheral receptors.

Any event that sets endolymph into motion can stimulate the equilibrium receptors. Placing an ice pack in contact with the temporal bone or flushing the external auditory canal with cold water may chill the endolymph in the outermost portions of the labyrinth and establish a temperature-related circulation of fluid. A mild and temporary vertigo is the result. Consumption of excessive quantities of alcohol and exposure to certain drugs can also produce vertigo by changing the composition of the endolymph or disturbing the hair cells.

In **Ménière's disease**, distortion of the membranous labyrinth by high fluid pressures may rupture the membranous wall and mix endolymph and perilymph together. The receptors in the vestibule and semicircular canals then become highly stimulated, and the individual may be unable to start a voluntary movement because of the intense spinning or rolling sensations experienced. In addition to the vertigo, the victim may "hear" unusual sounds as the cochlear receptors are activated. Other causes of vertigo include viral infection of the vestibular nerve and damage to the vestibular nucleus or its tracts.

The exceedingly unpleasant symptoms of **motion sickness** include headache, sweating, flushing of the face, nausea, vomiting, and various changes in mental perspective. (Sufferers may go from a state of giddy excitement to almost suicidal despair in a matter of moments.) It has been suggested that the condition results when central processing stations, such as the mesencephalic tectum, receive conflicting sensory information. Sitting belowdecks on a moving boat or reading a magazine in a car or airplane often provides the necessary conditions. Your eyes report that your position in space is not changing, but your labyrinthine receptors detect every bump and roll. As a result, "seasick" sailors watch the horizon rather than their immediate surroundings, so that the eyes will provide visual confirmation of the movements detected by the inner ear. It is not known why some individuals are almost immune to motion sickness, whereas others find travel by boat or plane almost impossible.

Drugs often administered to prevent motion sickness include *dimenhydrinate* (*Dramamine*®), *scopolamine*, and *promethazine*. These compounds appear to depress activity at the vestibular nuclei. Sedatives, such as *prochlorperazine* (*Compazine*®), may also be effective. In terms of convenience, the drug scopolamine is probably the easiest to use. It can be administered across the skin surface using an adhesive patch (*Transderm-Scop*™), and a single patch works for up to 72 hours.

 Testing and Treating Hearing Deficits *Page 546*

In the most common hearing test, a subject listens to sounds of varying frequency and intensity generated at irregular intervals. A record is kept of the responses, and the graphed record, or **audiogram**, is compared with that of an individual

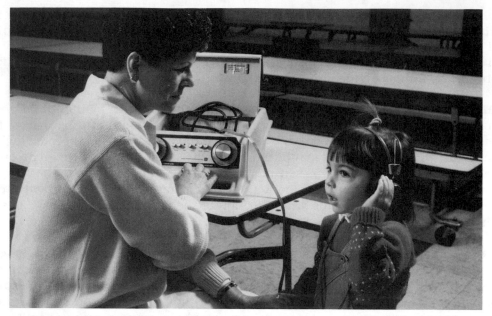

FIGURE 17-A
A hearing test

with "normal" hearing (see Figure 17-A). **Bone conduction** tests are used to discriminate between conductive and nerve deafness. If you put your fingers in your ears and talk quietly, you can still hear yourself because the bones of the skull conduct the sound waves to the cochlea, bypassing the middle ear. In a bone conduction test the physician places a vibrating tuning fork against the skull. If the subject hears the sound of the tuning fork in contact with the skull, but not when held next to the auditory meatus, the problem must lie within the external or middle ear. If the subject remains unresponsive to either stimulus, the problem must be at the receptors or along the auditory pathway.

Several effective treatments exist for conductive deafness. A hearing aid overcomes the loss in sensitivity by simply increasing the intensity of stimulation. Surgery may repair the tympanic membrane or free damaged or immobilized ossicles. Artificial ossicles may also be implanted if the originals are damaged beyond repair.

There are few possible treatments for nerve deafness. Mild conditions may be overcome by the use of a hearing aid if some functional hair cells remain. In a **cochlear implant** a small battery-powered device is inserted beneath the skin behind the mastoid process. Small wires run through the round window to reach the cochlear nerve, and when the implant "hears" a sound it stimulates the nerve directly. Increasing the number of wires and varying their implantation sites make it possible to create a number of different frequency sensations. Those sensations do not approximate normal hearing because there is as yet no way to target the specific afferent fibers responsible for the perception of a particular sound. Instead, a ran-

46

dom assortment of afferent fibers are stimulated, and the individual must learn to recognize the meaning and probable origin of the perceived sound.

Conjunctivitis

Page 550

Conjunctivitis, or "pink-eye," results from damage to and irritation of the conjunctival surface. The most obvious symptom results from dilation of the blood vessels beneath the conjunctival epithelium. The term *conjunctivitis* is more useful as the description of a symptom than as a name for a specific disease. A great variety of pathogens, including bacteria, viruses, and fungi, can cause conjunctivitis, and a temporary form of the condition may be produced by chemical or physical irritation (including even such mundane experiences as prolonged crying or peeling an onion).

Chronic conjunctivitis, or **trachoma**, results from bacterial or viral invasion of the conjunctiva. Many of these infections are highly contagious, and severe cases may disrupt the corneal surface and affect vision. The pathogen most often involved is *Chlamydia trachomatis*. Trachoma is a relatively common problem in southwestern North America, North Africa, and the Middle East. The condition must be treated with topical and systemic antibiotics to prevent scleral damage, eventual corneal damage, and vision loss.

Corneal Transplants

Page 552

The cornea has a very restricted ability to repair itself, so corneal injuries must be treated immediately to prevent serious visual losses. To restore vision after corneal scarring it is usually necessary to replace the cornea through a **corneal transplant**. Corneal replacement is probably the most common form of transplant surgery. Corneal transplants can be performed between unrelated individuals because there are no corneal blood vessels, and white blood cells that would otherwise reject the graft are unlikely to enter the area. Corneal grafts are obtained by posthumous donation; for best results the tissues must be removed within 5 hours after the donor's death.

Glaucoma affects roughly 2 percent of the population over 40. In this condition the aqueous humor no longer has free access to the canal of Schlemm. The primary factors responsible cannot be determined in 90 percent of all cases. Although drainage has been compromised, production of aqueous humor continues unabated, and the intraocular pressure begins to rise. The Fibrous scleral coat cannot expand significantly, so the increasing pressures begin to push against the surrounding soft tissues. The optic nerve is not wrapped in connective tissue, for it penetrates all three tunics. When intraocular pressures have risen to roughly twice normal

FIGURE 17-B
Using a tonometer to measure intraoccular pressure

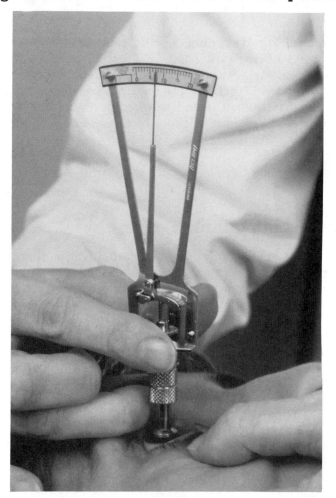

levels, distortion of the nerve fibers begins to affect visual perception. If this condition is not corrected, blindness eventually results.

Most eye exams include a glaucoma test. Until recently the standard procedure was to apply a topical anesthetic to the eye and place a small pressure monitor, a **tonometer**, on the ocular surface (see Figure 17-B). A newer procedure tests intraocular pressures by bouncing a tiny blast of air off the surface of the eye and measuring the deflection produced. Glaucoma may be treated by the application of drugs that constrict the pupil. As the pupillary constrictor muscle contracts, it stretches the epithelium covering the iris and tenses the peripheral epithelium. The tension at the edge of the iris appears to make the surface more permeable to aqueous humor. The rate of diffusion goes up, and as aqueous humor enters the canal of Schlemm the pressure inside the eye decreases. Surgical correction involves perforating the wall of the anterior chamber to encourage drainage. Traditional surgery requires hospitalization, but laser surgery can be performed on an outpatient basis. A series of tiny holes are burned through the ciliary margin with a precisely directed laser beam. Because laser light travels through the transparent cornea without affecting it, the eye remains intact. Bleeding is minimal, for the heat of the laser beam seals any damaged vessels.

Night Blindness *Page 560*

The visual pigments of the photoreceptors are synthesized from vitamin A. The body contains vitamin A reserves sufficient for several months. If dietary sources are inadequate, these reserves are gradually exhausted, and the amount of visual pigment in the photoreceptors begins to decline. Daylight vision is affected, but during the day the light is usually bright enough to stimulate whatever visual pigments remain within the densely packed cone population. As a result, the problem first becomes apparent at night, when the dim light proves insufficient to activate the rods. This condition, known as **night blindness**, can be treated by administration of vitamin A. The **carotene** pigments found in many vegetables can be converted to vitamin A within the body. Carrots are a particularly good source of carotene, which explains the old saying that carrots are good for your eyes.

Scotomas and Floaters *Page 562*

Abnormal blind spots, or **scotomas**, that are fixed in position may result from compression of the optic nerve, damage to photoreceptors, or central damage along the visual pathway. Scotomas are abnormal, permanent features of the visual field. Most readers will probably be more familiar with *floaters*, small spots that drift across the field of vision. Floaters are common, temporary phenomena that result from blood cells or cellular debris within the vitreous body. They can be detected by staring at a blank wall or a white sheet of paper.

Analyzing Sensory Disorders   *Page 565*

A recurring theme of this text is that an understanding of how a system works enables you to predict how things might go wrong. You are already familiar with the organization and physiology of sensory systems, and some of the most important clinical problems were discussed in clinical comments on the preceding pages. Placing the entire array into categories provides an excellent example of a strategy that can be used to analyze any system in the body.

Every sensory system contains peripheral receptors, afferent fibers, ascending tracts, nuclei, and areas of the cerebral cortex. Any malfunction affecting the system

FIGURE 17-C
Diagnosing sensory disorders

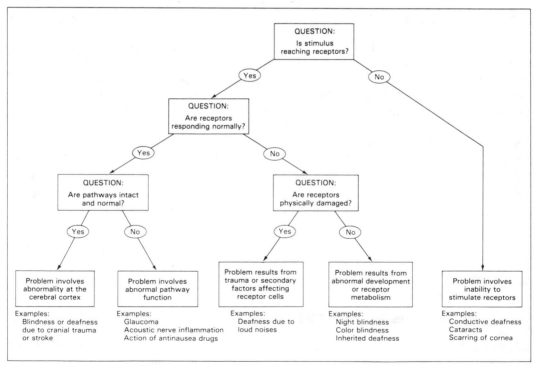

must involve one of those components. Any clinical diagnosis requires seeking answers to a series of yes or no questions, eliminating one possibility at a time until the nature of the problem becomes apparent. Figure 17-C organizes the disorders considered in this chapter into a "trouble-shooting" format similar to that used to diagnose problems with automobiles or other mechanical devices.

CHAPTER 18
The Endocrine System

Thyroid Gland Disorders   *Page 583*

Normal production of thyroid hormones establishes the background rates of cellular metabolism. These hormones exert their primary effects on large, metabolically active tissues and organs, including skeletal muscles, the liver, and the kidneys. Inadequate production of thyroid hormones, or **hypothyroidism**, in an infant is marked by inadequate skeletal and nervous development and a metabolic rate as much as 40 percent below normal levels. This condition affects approximately 1 birth out of every 5000. Hypothyroidism later in childhood will retard growth and delay puberty. Adults with this condition are lethargic and unable to adjust to cold temperatures. The symptoms, collectively known as **myxedema** (miks-e-DĒ-ma), include subcutaneous swelling, dry skin, hair loss, low body temperature, muscular weakness, and slowed reflexes.

Hypothyroidism usually results from some problem involving the thyroid gland rather than with pituitary production of TSH. One useful testing procedure involves monitoring the rate of iodine uptake by the thyroid gland, which provides an indication of its functional capabilities. Thyroid hormone and TSH levels may also be analyzed. Treatment involves the administration of synthetic thyroid hormones to maintain normal blood concentrations.

A **goiter** is an enlargement of the thyroid gland. That enlargement may not necessarily indicate increased production of thyroid hormones, merely an increase in follicular size. One form of goiter occurs if the thyroid fails to obtain enough iodine to meet its synthetic requirements. Under TSH stimulation, follicle cells produce large quantities of thyroglobulin, but they are unable to provide the iodine needed to create functional thyroid hormones. As a result, TSH levels rise further, and the thyroid gland begins to enlarge. Administering iodine may not solve the problem, for the sudden availability of iodine may produce symptoms of hyperthyroidism

51

(see below) as the stored thyroglobulin becomes activated. The usual therapy involves the injection of thyroxine, which feeds back on the hypothalamus and pituitary to inhibit the production of TSH. Over time the resting thyroid may return to its normal size and functional capabilities. **Thyrotoxicosis**, or **hyperthyroidism**, occurs when thyroid hormones are produced in excessive quantities. The metabolic rate climbs, and the skin becomes flushed and moist with perspiration. Blood pressure and heart rate increase, and the heartbeat may become irregular as circulatory demands escalate. The effects on the CNS make the individual restless, excitable, and subject to shifts in mood and emotional states. Despite the drive for increased activity, the subject has limited energy reserves and fatigues easily.

In **Graves' disease** excessive thyroid activity leads to goiter and the symptoms of hyperthyroidism. Protrusion of the eyes, or **exophthalmos** (eks-ahf-THAL-mos) may also appear, for unknown reasons. Graves' disease has a genetic basis and affects women much more often than men. Treatment may involve the use of antithyroid drugs, surgical removal of portions of the glandular mass, or destruction of part of the gland by exposure to radioactive iodine.

Hyperthyroidism may also result from thyroid tumors, inflammation, or immune system disorders. In extreme cases the individual's metabolic processes accelerate out of control. During a **thyrotoxic crisis** the subject experiences an extremely high fever, rapid heart rate, and the malfunctioning of a variety of physiological systems.

Disorders of Parathyroid Function   *Page 585*

When the parathyroid gland secretes inadequate or excessive amounts of parathormone, calcium concentrations move outside of normal homeostatic limits. Inadequate parathormone production, a condition called **hypoparathyroidism**, leads to low calcium concentrations in body fluids. The most obvious symptoms involve neural and muscle tissues, where calcium ions have important functions. The nervous system becomes more excitable, and the affected individual may experience **hypocalcemic tetany**, characterized by spasms in the muscles of the arms, hands, and face. Hypoparathyroidism may develop after neck surgery if the blood supply to these glands is restricted. In many other cases the primary cause of the condition is uncertain. Treatment is difficult because at present PTH can be obtained only by extraction from the blood of normal individuals. Thus PTH is extremely costly, and because supplies are very limited PTH administration is not used to treat this condition, despite its probable effectiveness. As an alternative, a dietary combination of vitamin D and calcium can be used to elevate body fluid calcium concentrations. (As noted in Chapter 7 of the text, vitamin D stimulates the absorption of calcium ions across the lining of the digestive tract.)

In **hyperparathyroidism** calcium concentrations become abnormally high. Calcium salts in the skeleton are mobilized, and bones are weakened. On X-rays the bones have a light, airy appearance because the dense calcium salts no longer dominate the tissue. CNS function is depressed, thinking slows, memory is impaired, and the individual often experiences emotional swings and depression. Nausea and vomiting occur, and in severe cases the patient may become comatose. Muscle function deteriorates, and skeletal muscles become weak. Other tissues are often affected as calcium salts crystallize in joints, tendons, and the dermis, and calcium deposits may produce masses, called *kidney stones*, that block filtration and conduction passages in the kidney. Hyperparathyroidism most often results from a tumor of the parathyroid gland. Treatment involves the surgical removal of the cancerous tissue. Fortunately there are four parathyroids, and the secretion of even a portion of one gland can maintain normal calcium concentrations.

Disorders of the Adrenal Cortex   *Page 587*

Clinical problems related to the adrenal gland vary depending on which of the adrenal zones becomes involved. The conditions may result from changes in the functional capabilities of the adrenal cells (primary conditions) or disorders affecting the regulatory mechanisms (secondary conditions). In **hypoaldosteronism** the zona glomerulosa fails to produce enough aldosterone, usually because the kidneys are not releasing adequate amounts of renin. Affected individuals lose excessive amounts of water and sodium ions at the kidneys, and the water loss leads to low blood volume and pressure. Changes in electrolyte concentrations affect transmembrane potentials, eventually causing dysfunctions in neural and muscular tissues.

Hypersecretion of aldosterone results in the condition of **aldosteronism**. Under continued aldosterone stimulation, the kidneys retain sodium ions very effectively, but potassium ions are lost in large quantities. In response, potassium ions move out of the cells and into the interstitial fluids, only to be lost in turn. A crisis eventually develops when low extracellular potassium concentrations disrupt normal cardiac, neural, and kidney cell functions.

Addison's disease may result from inadequate stimulation of the zona fasciculata by ACTH, or (rarely) from the inability of the cells to synthesize the necessary hormones. Affected individuals become weak and lose weight, due to a combination of appetite loss and digestive tract malfunctions. They cannot adequately mobilize energy reserves, and their blood glucose concentrations fall sharply within hours after a meal. Stresses cannot be tolerated, and a minor infection or injury may lead to a sharp and fatal decline in blood pressure. A particularly interesting symptom is the increased melanin pigmentation in the skin. The ACTH molecule and the MSH molecule are similar in structure, and at high concentrations ACTH stimulates the MSH receptors on melanocytes.

Cushing's disease results from overproduction of glucocorticoids. The symptoms resemble those of a protracted and exaggerated response to stress. (The stress response is discussed later in this chapter.) Glucose metabolism is suppressed, lipid reserves are mobilized, and peripheral proteins are broken down. Lipids and amino acids are mobilized in excess of the existing demand. The energy reserves are shuffled around, and the distribution of body fat changes. Adipose tissues in the cheeks and around the base of the neck become enlarged at the expense of other areas, producing a "moon-faced" appearance. The demand for amino acids falls most heavily on the skeletal muscles, which respond by breaking down their contractile proteins. This response reduces muscular power and endurance.

The chronic administration of large doses of steroids can produce symptoms similar to those of Cushing's disease, but such treatment is usually avoided. Roughly 75 percent of cases result from an overproduction of ACTH, and afflicted individuals may also show changes in skin pigmentation. If the problem stems from a tumor within the adrenal cortex, the zona reticularis may also be affected. The overproduction of androgens then produces symptoms of the **androgenital syndrome**. In women this condition leads to the gradual development of male secondary sexual characteristics, including body and facial hair patterns, adipose tissue distribution, and muscular development. Tumors affecting the zona reticularis of males may sometimes result in the production of large quantities of estrogens. In this condition, **gynecomastia** (*gynaikos*, woman + *mastos*, breast), the individual develops female secondary sexual characteristics.

⚕ Disorders of the Adrenal Medulla

Page 588

The overproduction of epinephrine by the adrenal medulla may reflect chronic sympathetic activation. A **pheochromocytoma** (fē-ō-krō-mō-sī-TŌ-mah) is a tumor that produces catecholamines in massive quantities. The tumor usually develops within the adrenal medulla, but it may also involve other sympathetic ganglia. The most dangerous symptoms are rapid and irregular heartbeat and high blood pressure; other symptoms include uneasiness, sweating, blurred vision, and headaches. This condition is rare, and surgical removal of the tumor is the most effective treatment.

CHAPTER 19
The Cardiovascular System: The Blood

Bilirubin Tests and Jaundice   *Page 613*

Normal plasma bilirubin concentrations range from 0.5 to 1.2 mg/dℓ (mg per 100 mℓ). Of that amount, roughly 85 percent is unconjugated bilirubin on its way to the liver. Several different clinical conditions are characterized by an increase in the total plasma bilirubin concentration. In such conditions, bilirubin diffuses into peripheral tissues, giving them a yellow coloration that is most apparent in the skin and over the sclera of the eyes. This combination of signs (yellow skin and eyes) is called **jaundice** (JAWN-dis).

Jaundice can have many different causes, but blood tests that determine the concentration of unconjugated and conjugated forms of bilirubin can provide useful diagnostic clues. For example, **hemolytic jaundice** results from the destruction of large numbers of red blood cells. When this occurs, phagocytes release massive quantities of unconjugated bilirubin into the blood. Because the liver cells accelerate the secretion of bilirubin in the bile, the blood concentration of conjugated bilirubin does not increase proportionately. So a blood test from a patient with hemolytic jaundice would reveal: (1) elevated total bilirubin, (2) high concentrations of unconjugated bilirubin, and (3) conjugated bilirubin contributing much less than 15 percent to the total bilirubin concentration.

These results are quite different from those seen in **obstructive jaundice**. In this condition, discussed in Chapter 24 of the text, the ducts that remove bile from the liver are constricted or blocked. Liver cells cannot get rid of conjugated bilirubin, and large quantities diffuse into the blood. In this case diagnostic tests would show: (1) elevated total bilirubin, (2) unconjugated bilirubin contributing much less than 85 percent to the total bilirubin concentration, and (3) high concentrations of conjugated bilirubin.

The Leukemias

Leukocytosis with white blood cell counts of 100,000/μl or more usually indicate the presence of some form of **leukemia** (loo-KĒ-mē-ah). Leukemias characterized by the presence of abnormal granulocytes or other cells of the bone marrow are called **myeloid**, and those involving lymphocytes are termed **lymphoid**. The first symptoms appear as immature and abnormal white blood cells appear in the circulation. As their numbers increase they travel via the circulation, invading tissues and organs throughout the body.

These cells are extremely active, and they require abnormally large amounts of energy. As in other cancers, described in Chapter 5 of the text (see the Clinical Comment on p. 163), invading leukemic cells gradually replace the normal cells, especially in the bone marrow. Red blood cell and platelet formation decline, with resulting anemia and impaired blood clotting, and untreated leukemias are invariably fatal. Leukemias may be classified as acute (short and severe) or chronic (prolonged). Acute leukemias may be linked to radiation exposure, hereditary susceptibility, virus infections, or unknown causes. Chronic leukemias may be related to chromosomal abnormalities or immune system malfunctions. Survival in untreated acute leukemia averages about 3 months; individuals with chronic leukemia may survive untreated for years.

Effective treatments exist for some forms of leukemia and not others. For example, when acute lymphoid leukemia is detected early, 85–90 percent of patients can be held in remission for 5 years or longer, but only 10–15 percent of patients with acute myeloid leukemia survive 5 years or more. The yearly mortality rate for leukemia (all types) in the United States has not declined appreciably in the past 30 years, remaining at around 6.8 per 100,000 population.

One option for treating acute leukemias is to perform a **bone marrow transplant**. In this procedure massive chemotherapy or radiation treatment is given, enough to kill all of the cancerous cells. Unfortunately, this also destroys the patient's blood cells and stem cells in the bone marrow and other blood-forming tissues. The individual then receives an infusion of healthy bone marrow cells that repopulate the blood and marrow tissues.

If the bone marrow is extracted from another person (a *heterologous marrow transplant*), care must be taken to ensure that the blood types and tissue types are compatible (see Chapters 19 and 22 of the text). If they are not, the new lymphocytes may attack the patient's tissues, with potentially fatal results. Best results are obtained when the donor is a close relative. In an *autologous marrow transplant* bone marrow is removed from the patient, cleansed of cancer cells, and reintroduced after radiation or chemotherapy treatment. Although there are fewer complications, the preparation and cleansing of the marrow are technically difficult and time-consuming.

Bone marrow transplants are also performed to treat patients whose bone marrow has been destroyed by toxic chemicals or radiation. For example, heterologous transplants were used successfully in the U.S.S.R. to treat survivors of the Chernobyl nuclear reactor accident in 1986.

Disseminated Intravascular Coagulation   *Page 625*

The clotting process is complex, and normally is precisely regulated. In **disseminated intravascular coagulation (DIC)** bacterial toxins remove circulating fibrinogen by activating thrombin. The thrombin then converts fibrinogen to fibrin within the circulating blood. Much of the fibrin is removed by phagocytes or dissolved by plasmin, but small clots may block small vessels and damage peripheral tissues. If the liver cannot keep pace with the demand for fibrinogen, clotting abilities gradually decline, and uncontrolled bleeding may occur.

Hemophilia   *Page 626*

Hemophilia (hēm-ō-FĒL-ē-ah) is one of many inherited disorders characterized by inadequate production of clotting factors. The incidence of this condition in the general population is about 1 in 10,000, with males accounting for 80–90 percent of those affected. In hemophilia, production of a single clotting factor (most often Factor VIII) is reduced; the severity of the condition depends on the degree of reduction. In severe cases extensive bleeding accompanies the slightest mechanical stresses, and hemorrhages occur spontaneously at joints and around muscles.

Transfusions of clotting factors can often reduce or control the symptoms of hemophilia, but plasma samples from many individuals must be pooled (combined) to obtain adequate amounts of clotting factors. This makes the procedure very expensive, and increases the risk of infection with blood-borne infections such as hepatitis or AIDS. Gene splicing techniques have been used to manufacture clotting factor VIII, an essential component of the intrinsic clotting pathway. Although supplies are now limited, this procedure should eventually provide a safer and cheaper method of treatment.

Testing the Clotting System

Several clinical tests check the efficiency of the clotting system:

Bleeding time: This test measures the time it takes for a small puncture wound to seal itself. There are several variations on this procedure, with normal values ranging from 1 to 9 minutes.

Coagulation time: In this test, a sample of whole blood is allowed to stand under controlled conditions until a visible clot has formed. Normal values range from 3 to 15 minutes. The test has several potential sources of error, and so is not very accurate. It is nevertheless of value because it is the simplest test that can be performed on a blood sample. More sophisticated tests begin by adding citrate ions to the sample. Citrate ties up the calcium ions in the plasma and prevents premature clotting.

Partial thromboplastin time (**PTT**): In this test a plasma sample is mixed with chemicals that mimic the effects of activated platelets. Calcium ions are then introduced, and the clotting time is recorded. Clotting normally occurs in 35–50 seconds if the enzymes and clotting factors of the intrinsic pathway are present in normal concentrations.

Plasma prothrombin time (*prothrombin time*, **PT**): This test checks the performance of the extrinsic pathway. The procedure is similar to that in the PTT test, but the clotting process is triggered by exposure to a combination of tissue thromboplastin and calcium ions. Clotting normally occurs in 12–14 seconds.

CHAPTER 20

The Cardiovascular System: The Heart

Infection and Inflammation of the Heart *Page 640*

Many different microorganisms may infect heart tissue, leading to serious cardiac abnormalities. **Carditis** (kar-DĪ-tis) is a general term indicating inflammation of the heart. Clinical conditions resulting from cardiac infection are usually identified by the primary site of infection. For example, those affecting the endocardium produce symptoms of **endocarditis**. Endocarditis primarily affects the chordae tendineae and valves, and the mortality rate may reach 21–35 percent. The most severe complications result from the formation of blood clots on the damaged surfaces. These clots subsequently break free, entering the circulation as drifting emboli (see p. 627 of the text) that may cause strokes, heart attacks, or kidney failure.

Bacteria, viruses, protozoa, and fungal pathogens that attack the myocardium produce **myocarditis**. The microorganisms implicated include those responsible for many of the conditions discussed in earlier chapters, including diphtheria, syphilis, polio, and malaria. The infected heart muscle becomes extremely sensitive, and the heart rate rises dramatically. Over time, abnormal contractions may appear, and these may prove fatal.

An equally great variety of pathogens can infect the pericardium, producing a condition called **pericarditis**. The inflamed pericardial layers rub against one another as the heart beats, producing a characteristic scratching sound. In addition, pericardial irritation and inflammation often result in an increased production of pericardial fluid. Fluid then collects in the pericardial sac, potentially restricting the movement of the heart. This condition is called **cardiac tamponade** (tam-po-NĀD).

Cardiac Contractions and the Composition of the Extracellular Fluid

Page 643

Alterations in extracellular calcium and potassium ion concentrations have particularly dramatic effects on cardiac function. Variations in the calcium ion concentration primarily affect the force of cardiac contraction. If the extracellular concentration of calcium is elevated, a condition termed **hypercalcemia** (hī-per-kal-SĒ-me-a), cardiac muscle cells become extremely excitable, and their contractions become powerful and prolonged. In extreme cases the heart goes into an extended state of contraction that is usually fatal. When the calcium ion concentration is abnormally low, a condition termed **hypocalcemia** (hī-pō-kal-SĒ-mē-a), the contractions become very weak, and may cease altogether.

Changes in the potassium ion concentration primarily affect the transmembrane potential and the rates of depolarization and repolarization. When potassium ion concentrations are high, a condition termed **hyperkalemia** (hī-per-ka-LĒ-mē-a) exists. Muscle fiber membranes depolarize because the rate of potassium loss by the cell declines. In addition, repolarization becomes difficult because there is less of a concentration gradient to drive potassium ions through the voltage-gated channels and out of the muscle fiber. Under these circumstances, cardiac contractions become weak and irregular. When potassium ion concentrations are abnormally low, a condition termed **hypokalemia** (hī-pō-ka-LĒ-mē-a), the membranes of cardiac muscle fibers hyperpolarize, and their rate of contraction declines. Severe hyperkalemia and hypokalemia are life-threatening conditions that require immediate corrective action.

Problems with Pacemaker Function

Page 650

Symptoms of severe bradycardia (below 50 beats per minute) include weakness, fatigue, confusion, and loss of consciousness. Drug therapies are seldom helpful, but artificial pacemakers can be used with considerable success. Wires run to the atria, the ventricles, or both, depending on the nature of the problem, and the unit delivers small electrical pulses to stimulate the myocardium. Internal pacemakers are surgically implanted, batteries and all. These units last 7–8 years or more before another operation is required to change the battery. External pacemakers are used for temporary emergencies, such as immediately after cardiac surgery. Only the wires are implanted, and an external control box is worn on the belt.

There are over 50,000 artificial pacemakers in use at present. The simplest provide constant stimulation to the ventricles at rates of 70–80 per minute. More sophisticated pacemakers vary their rates to adjust to changing circulatory demands, as during exercise. Others are able to monitor cardiac activity and respond whenever the heart begins to function abnormally.

Tachycardia, usually defined as a heart rate of over 100 beats per minute, increases the workload on the heart. At very high heart rates cardiac performance suffers because the ventricles do not have enough time to refill with blood before the next contraction occurs. Chronic or acute incidents of tachycardia may be controlled by drugs that affect the permeability of pacemaker membranes or block the effects of sympathetic stimulation.

Valvular Heart Disease   *Page 656*

Abnormalities in AV or semilunar valve function reduce the efficiency of the heart. Minor valve problems frequently go unnoticed, and some degree of regurgitation often occurs in otherwise normal individuals. When valve function deteriorates to the point that the heart cannot maintain adequate circulatory flow, symptoms of **valvular heart disease** appear. Congenital malformations may be responsible, but often the condition develops after **carditis** (see the discussion of Infection and Inflammation in the Clinical Manual.

One common cause of carditis is **rheumatic** (roo-MA-tik) **fever**, an inflammatory condition that may develop following infection by streptococcal bacteria. Rheumatic fever most often affects children of ages 5–15; symptoms include high fever, joint pain and stiffness, and a distinctive full-body rash. Obvious symptoms usually persist for less than 6 weeks, although severe cases may linger for 6 months or more. The longer the duration of the inflammation, the more likely it is that carditis will develop. The carditis that does develop in 50–60 percent of patients often escapes detection, and scar tissue forms gradually in the myocardium and the heart valves. Valve condition deteriorates over time, and valve problems serious enough to affect cardiac function may not appear until 10–20 years after the initial infection.

Over the interim the affected valves become thickened and often calcified to some degree. This thickening narrows the opening guarded by the valves, producing a condition called **valvular stenosis** (ste-NŌ-sis; *stenos*, narrow). The resulting clinical disorder is known as **rheumatic heart disease**, or **RHD**. The thickened cusps stiffen in a partially closed position, but the valves do not completely block the circulation because the edges of the cusps are rough and irregular. Regurgitation may occur, and much of the blood pumped out of the heart may flow right back in.

Mitral stenosis and **aortic stenosis** are the most common forms of valvular heart disease. About 40 percent of patients with RHD develop mitral stenosis, and two-thirds of them are women. The reason for the correlation between female gender and mitral stenosis is unknown. In mitral stenosis blood enters the left ventricle at a slower than normal rate, and when the ventricle contracts blood flows back into the left atrium (mitral regurgitation) as well as into the aortic trunk. As a result, the left ventricle has to work much harder to maintain adequate systemic circulation. The right and left ventricles discharge identical amounts of blood with each beat, and as the output of the left ventricle declines blood "backs up" in the pulmonary circuit. Venous pressures then rise in the pulmonary circuit, and the right ventricle must develop greater pressures to force blood into the pulmonary trunk. In severe cases of mitral stenosis the ventricular musculature is not up to the task. The heart weakens, and peripheral tissues begin to suffer from oxygen and nutrient deprivation. (This condition, called **heart failure**, is discussed in more detail in Chapter 21 of this Manual.)

Symptoms of aortic stenosis develop in roughly 25 percent of patients with RHD; 80 percent of these individuals are males. Symptoms of aortic stenosis are initially less severe than those of mitral stenosis. Although the left ventricle enlarges and works harder, normal circulatory function can often be maintained for years. Clinical problems develop only when the opening narrows enough to prevent adequate blood flow. Symptoms then resemble those of mitral stenosis.

One reasonably successful treatment for severe stenosis involves the replacement of the damaged valve with a prosthetic (artificial) valve. Figure 20-A shows a stenotic

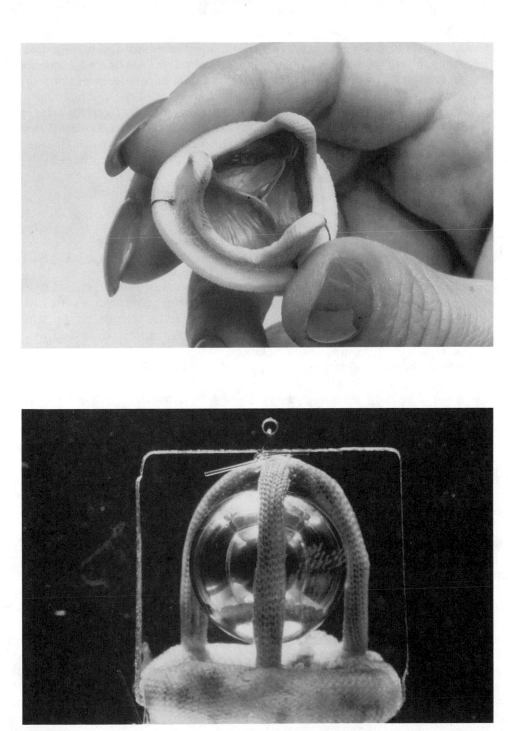

FIGURE 20-A
Stenotic human heart valve (bottom page 62), pig heart valve and synthetic valve (above)

heart valve (bottom of page 62) and two possible replacements: a valve from a pig (top of page 63) and a synthetic valve (bottom of page 63), one of a number of designs that have been employed. The plastic and/or stainless steel components of the artificial valve do not activate the immune system of the recipient, and the smooth surfaces reduce complications because they are less likely to trigger the clotting system. (Oral anticoagulant therapy is still necessary, however.) Valve replacement operations are quite successful, with about 95 percent of the surgical patients surviving for three years or more, and 70 percent surviving over five years.

The Cardiomyopathies *Pages 648, 654*

The **cardiomyopathies** (kar-dē-ō-mī-OP-a-thēz) include an assortment of diseases with a common symptom: the progressive, irreversible degeneration of the myocardium. Cardiac muscle fibers are damaged and replaced by fibrous tissue, and the muscular walls of the heart become thin and weak. As muscle tone declines, the ventricular chambers become greatly enlarged. When the remaining fibers cannot develop enough force to maintain cardiac output, symptoms of heart failure develop.

Chronic alcoholism and coronary artery disease are probably the most common causes of cardiomyopathy in the United States. Infectious agents, including viruses, bacteria, fungi, and protozoans, can also produce cardiomyopathies. Diseases affecting neuromuscular performance, such as muscular dystrophy (see Chapter 10 of this Manual), can also damage cardiac muscle fibers, as can starvation or chronic variations in the extracellular concentrations of calcium or potassium ions. Finally there are several inherited forms of cardiomyopathy, as well as a significant number of cases in which the primary cause cannot be determined.

Individuals suffering from severe cardiomyopathies may be considered as candidates for heart transplants. This surgery involves the complete removal of the weakened heart and its replacement by a heart taken from a suitable donor. To survive the surgery, the recipient must be in otherwise satisfactory health. Because the number of suitable donors is limited, the available hearts are usually assigned to individuals younger than age 50. Out of the 8000–10,000 U.S. patients each year suffering from potentially fatal cardiomyopathies, only around 1000 receive heart transplants. There is an 80–85 percent 1-year survival rate, and a 50–70 percent 5-year survival rate after successful transplantation. This rate is quite good, considering that these patients would have died if the transplant had not been performed.

Unfortunately, many individuals with cardiomyopathy who are initially selected for surgery succumb to the disease before a suitable donor becomes available. For this reason there continues to be considerable interest in the development of an artificial heart. One model, the Jarvik-7, had limited clinical use in the 1980s.

64

Attempts to implant it on a permanent basis were unsuccessful, primarily because of formation of blood clots on the mechanical valves. When these clots broke free, they formed drifting emboli that plugged peripheral vessels, producing strokes, kidney failure, and other complications. In 1989 the federal government prohibited further experimental use of the Jarvik-7 as a permanent heart implant. Modified versions of this unit and others now under development may still be used to maintain transplant candidates while awaiting the arrival of a donor organ.

An interesting experimental procedure involves using skeletal muscle tissue to apply permanent patches to injured hearts or to build small accessory pumps. For example, one procedure involves freeing a portion of the latissimus dorsi muscle from the side. This flap, with its circulation intact, is moved into the thoracic cavity and folded to form a sling around the heart. A pacemaker is then used to stimulate its contraction, and when it contracts it squeezes the heart and helps push blood into the major arteries. These methods are less stressful than heart transplants because (1) they leave the damaged heart in place, and (2) the transplanted tissue is taken from the same individual, so it will not be attacked by the immune system. Although preliminary results have been encouraging, few procedures have been performed, and the use of skeletal muscle patches remains an experimental concept rather than a recognized treatment for cardiomyopathy.

 Interpreting Abnormal ECGs   *Page 657*

Conduction deficits   Damage to the conduction pathways caused by mechanical distortion, ischemia, infection, or inflammation can affect the normal rhythm of the heart. The resulting condition is called a **conduction deficit**, or **heart block**. Heart blocks of varying severity are illustrated in Figure 20-B. In a **first-degree heart block** (Figure 20-B2) the AV node and proximal portion of the AV bundle slow the passage of impulses heading for the ventricular myocardium. As a result, a pause appears between the atrial and ventricular contractions. Although a delay exists, the regular rhythm of the heart continues, and each atrial beat is followed by a ventricular contraction.

If the delay lasts long enough, the nodal cells will still be repolarizing from the previous beat when the next impulse arrives from the pacemaker. The arriving impulse will then be ignored, the ventricles will not be stimulated, and the normal "atria-ventricles, atria-ventricles" pattern will disappear. This condition is a **second-degree heart block** (Figure 20-B3). A mild second-degree block may produce only an occasional skipped beat, but with more substantial delays the ventricles will follow every second atrial beat. The resulting pattern of "atria, atria-ventricles, atria, atria-ventricles" is known as a **two-to-one** (2:1) **block. Three-to-one** or even **four-to-one** blocks are also encountered.

65

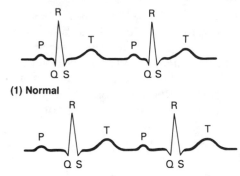

(1) Normal

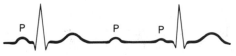

(2) First-degree heart block (long P–R internal)

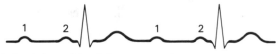

Skipped ventricular beat

2:1 Block (ventricles follow every other atrial beat)

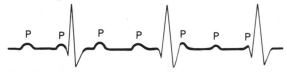

3:1 Block (ventricles follow every third atrial beat)

(3) Second-degree blocks

(4) Complete block or third-degree block
(atrial beats occur regularly, ventricular beats occur at slower, unrelated pace)

FIGURE 20-B
Heart blocks

In a **third-degree**, or **complete**, **heart block** the conducting pathway stops functioning altogether (Figure 20-B4). The atria and ventricles continue to beat, but their activities are no longer synchronized. The atria follow the pace set by the SA node, beating 70–80 times per minute, and the ventricles follow the commands of the AV node, beating at a rate of 40–60 per minute. A temporary third-degree block can be induced by stimulating the vagus nerve. In addition to slowing the rate of impulse generation by the SA node, such stimulation inhibits the AV nodal cells

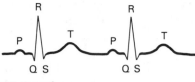

(1) Normal

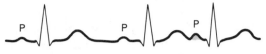

(2) Premature atrial contraction (PAC)

(3) Paroxysmal atrial tachycardia (PAT)

(4) Atrial fibrillation

(5) Premature ventricular contraction (PVC)

(6) Ventricular tachycardia (VT)

(7) Ventricular fibrillation (VF)

FIGURE 20-C
Abnormal atrial or ventricular function

to the point that they cannot respond to normal stimulation. Comments such as "my heart stopped," or "my heart skipped a beat" usually refer to this phenomenon. The pause typically lasts for just a few seconds. Longer delays end when a conducting cell, usually one of the Purkinje cells, depolarizes to threshold. This phenomenon is called **ventricular escape** because the ventricles are escaping from the control

67

of the SA node. Ventricular escape can be a lifesaving event if the conduction system is damaged. Even without instructions from the SA or AV nodes, the ventricles will continue to pump blood at a slow but steady rate.

Abnormal atrial or ventricular function Other important examples of arrhythmias are shown in Figure 20-C. **Premature atrial contractions** (**PACs**), indicated in 20-C2, often occur in normal individuals. In a PAC the normal atrial rhythm is momentarily interrupted by a "surprise" atrial contraction. Stress, caffeine, and various drugs may increase the frequency of PAC incidence, presumably by increasing the permeabilities of the SA pacemakers. The impulse spreads along the conduction pathway, and a normal ventricular contraction follows the atrial beat.

In **paroxysmal atrial tachycardia** (par-ok-SIZ-mal; "sudden attack"), or **PAT** (Figure 20-C3), a premature atrial contraction triggers a flurry of atrial activity. The ventricles are still able to keep pace, and the heart rate jumps to about 180 beats per minute. In **atrial flutter** the atria are contracting in a coordinated manner, but the contractions are occurring very frequently. During a bout of **atrial fibrillation** (fi-bri-LĀ-shun), Figure 20-C4, the impulses are moving over the atrial surface at rates of perhaps 500 beats per minute. The atrial wall quivers instead of producing an organized contraction. Despite the fact that the atria are now essentially nonfunctional, the condition may go unnoticed, especially in older individuals leading sedentary lives. PACs, PAT, atrial flutter, and even atrial fibrillation are not considered dangerous unless they are associated with some more serious indications of cardiac damage, such as coronary artery disease or valve problems.

In contrast, ventricular arrhythmias are serious and often fatal. Because the conduction system functions in one direction only, a ventricular arrhythmia is not linked to atrial activities. **Premature ventricular contractions** (**PVCs**; Figure 20-C5) occur when a Purkinje cell or ventricular myocardial cell depolarizes to threshold and triggers a premature contraction. The cell responsible is called an **ectopic pacemaker** (ek-TOP-ik; "out of place"). The frequency of PVCs can be increased by exposure to epinephrine and other stimulatory drugs or to ionic changes that depolarize cardiac muscle fiber membranes. Similar factors may be responsible for periods of **ventricular tachycardia**, also known as **VT**, or "V-tach" (Figure 20-C6).

PVCs and VTs often precede the most serious arrhythmia, **ventricular fibrillation** (**VF**), shown in Figure 20-C7. The resulting condition, also known as **cardiac arrest**, is rapidly fatal because the heart stops pumping blood. During ventricular fibrillation the cardiac muscle fibers are overly sensitive to stimulation, and the impulses are traveling from cell to cell around and around the ventricular walls. A normal rhythm cannot become established because the ventricular muscle fibers are stimulating one another at such a rapid rate. A **defibrillator** is a device that attempts to eliminate ventricular fibrillation and restore normal cardiac rhythm. Two electrodes are placed in contact with the chest and a powerful electrical shock is administered. The electrical stimulus depolarizes the entire myocardium simultaneously.

With luck, after repolarization the SA node will be the first area of the heart to reach threshold. Thus the primary goal of defibrillation is not just to stop the fibrillation, but to give the ventricles a chance to respond to normal SA commands.

CHAPTER 21

The Cardiovascular System: Vessels and Circulation

Aneurysms

Page 664

An **aneurysm** (AN-ū-rizm) is a bulge in the weakened wall of a blood vessel, usually an artery. This bulge resembles a bubble in the wall of a tire, and like a bad tire, the affected artery may suffer a catastrophic blowout. The most dangerous aneurysms are those involving arteries of the brain, where they cause strokes, and of the aorta, where a blowout will cause fatal bleeding in a matter of seconds.

Aneurysms are most often caused by chronic high blood pressure. With age, the vessel walls become less elastic, and when a weak point develops the high arterial pressures distort the wall, creating an aneurysm. Unfortunately, because they are often painless, they are likely to go undetected. When aneurysms are found by ultrasound or other scanning procedures, the risk of rupture can sometimes be estimated on the basis of their size. For example, an aortic aneurysm larger than 6 cm has a 50:50 chance of rupturing in the next 10 years. Treatment often begins with the reduction of blood pressure by means of vasodilators or beta-blockers (see pp. 511 and 644 of the text). An aneurysm in an accessible area, such as the abdomen, may be surgically removed and the vessel repaired. Figure 21-A shows a large aortic aneurysm before and after surgical repair with a synthetic patch.

Although high blood pressure is most often responsible for aneurysm formation, any trauma or infection that weakens vessel walls can lead to an aneurysm. In

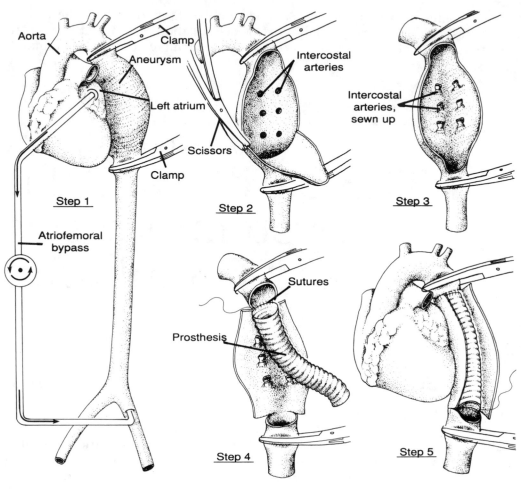

FIGURE 21-A
Repair of an aneurysm

addition, at least some aortic aneurysms have been linked to inherited disorders, such as Marfan's syndrome (discussed in Chapters 5 and 7 of this Manual) that weaken connective tissues. Affected individuals have weak arterial walls, making them more likely to develop aortic aneurysms. It is not known whether genetic factors are involved in the development of other aneurysms.

Problems with Venous Valve Function   *Page 672*

In Chapter 5 of the text it was noted that one of the consequences of aging is a loss of elasticity and resilience in connective tissues throughout the body. Blood vessels are no exception, and with age the walls of veins begin to sag. This change usually affects the superficial veins of the arms and legs first, because at these locations gravity opposes blood flow. The situation is aggravated by a lack of exercise or an occupation requiring long hours standing or sitting. Because there is no muscular activity to help keep the blood moving, venous blood pools on the proximal (heart) side of each valve. As the venous walls are distorted, the valves become less effective, and gravity can then pull blood back toward the capillaries. This further impedes normal blood flow, and the veins become grossly distended. These sagging, swollen vessels are called **varicose** (VAR-i-kōs) **veins**. Varicose veins are relatively harmless but unsightly; surgical procedures are sometimes used to remove or constrict the offending vessels.

Varicose veins are not limited to the extremities, and another common site involves a network of veins in the walls of the anus. Pressures within the abdominopelvic cavity rise dramatically when the abdominal muscles are tensed. Straining to force defecation can force blood into these veins, and repeated incidents leave them permanently distended. These distended veins, known as **hemorrhoids** (HEM-ō-roidz), can be uncomfortable and in severe cases extremely painful.

Hemorrhoids are often associated with pregnancy, due to changes in circulation and abdominal pressures. Minor cases can be treated by the topical application of drugs that promote contraction of smooth muscles within the venous walls. More severe cases may require the surgical removal or destruction of the distended veins.

Treatment of Cerebrovascular Disease   *Page 679*

Cerebrovascular disease was introduced in Chapter 15 of the text (see p. 492); this section considers additional details related to the concepts in this chapter. Symptoms of cerebrovascular disease usually appear when atherosclerosis reduces the circulatory supply to the brain. If the circulation to a portion of the brain is completely shut off, a cerebrovascular accident (CVA), or stroke, occurs. The most common causes of strokes include **cerebral thrombosis** (clot formation at a plaque), **cerebral embolism** (drifting blood clots, fatty masses, or air bubbles), and **cerebral hemorrhages** (rupture of a blood vessel, often following formation of an aneurysm). The observed symptoms and their severity vary depending on the vessel involved and the location of the blockage. (Examples were included in the discussion in Chapter 15 of the text.)

If the circulatory blockage disappears in a matter of minutes, the effects are temporary, and the condition is called a **transient ischemic attack**, or **TIA**. TIAs often indicate that cerebrovascular disease exists, and preventive measures can be taken to forestall more serious incidents. For example, taking aspirin each day slows blood clot formation in patients experiencing TIAs, and this reduces the risks of cerebral thrombosis and cerebral embolism.

If the blockage persists for a longer period, neurons die and the area degenerates. Stroke symptoms are initially exaggerated by the swelling and distortion of the injured neural tissues; if the individual survives, there is often a gradual improvement in brain function. The management and treatment of strokes remain controversial. Surgical removal of the offending clot or blood mass may be attempted, but the results are variable. Recent progress in the emergency treatment of cerebral thromboses and cerebral embolisms has involved the administration of clot-dissolving enzymes such as tissue plasminogen activator (t-PA; now sold as *Alteplase*), streptokinase, or urokinase. Best results are obtained if these enzymes are administered within an hour, although they may still be of use up to 24 hours after the stroke. Subsequent treatment involves anticoagulant therapy, usually with heparin (for 1–2 weeks) followed by coumadin (for up to a year) to prevent further clot formation. (These fibrinolytic and anticoagulant drugs were introduced in Chapter 19 of the text (see p. 627). A more complicated surgical procedure involves the insertion of a transplanted piece of a blood vessel that routes blood around the damaged area. None of these treatments is as successful as preventive surgery, where plaques are removed *before* a stroke. It should also be noted that the very best solution is to prevent or restrict plaque formation by controlling the risk factors involved.

Hypertension and Hypotension *Page 692*

The usual criterion for **hypertension** in an adult is a blood pressure greater than 150/90. One study estimated that 20 percent of the U.S. Caucasian population has blood pressures greater than 160/95 and that another 25 percent is on the borderline, with pressures over 140/90. The figures for other racial groups vary; the percentages for black Americans are even higher.

Elevated blood pressure is considered **primary hypertension**, or **essential hypertension**, if no obvious cause can be determined. Known risk factors include a hereditary history of hypertension, sex (males are at higher risk), high plasma cholesterol, obesity, chronic stresses, and cigarette smoking. **Secondary hypertension** appears as the result of abnormal hormonal production outside the cardiovascular system. For example, a condition resulting in excessive production of antidiuretic hormone (ADH), renin, aldosterone, or epinephrine will probably produce hypertension.

Hypertension significantly increases the workload on the heart, and the left ventricle gradually enlarges. More muscle mass requires a greater oxygen demand, and when the coronary circulation cannot keep pace symptoms of coronary ischemia appear.

Increased arterial pressures also place a physical stress on the walls of blood vessels throughout the body. This stress promotes or accelerates the development of arteriosclerosis and increases the risks of aneurysms, heart attacks, and strokes. Vessels supplying the retinas of the eyes are often affected, and hemorrhages and associated circulatory changes can produce disturbances in vision. Because these vessels are examined in a normal physical exam, retinal changes may provide the first evidence that hypertension is affecting peripheral circulation.

One of the most difficult aspects of hypertension is that there are usually no obvious symptoms. As a result, clinical problems do not appear until the condition has reached the crisis stage. There is therefore considerable interest in early detection and prompt treatment of hypertension.

Treatment consists of a combination of lifestyle changes and physiological therapies. Quitting smoking, getting regular exercise, and restricting dietary intake of salt, fats, and calories will improve peripheral circulation, prevent increases in blood volume and total body weight, and reduce plasma cholesterol levels. These strategies may be sufficient to control hypertension if it has been detected before significant cardiovascular damage has occurred. Therapies usually involve antihypertensive drugs, such as calcium channel blockers, beta-blockers, diuretics, and vasodilators, singly or in combination (see pp. 644–646 of the text). Beta-blockers eliminate the effects of sympathetic stimulation on the heart, and the unopposed parasympathetic system lowers the resting heart rate and blood pressure. Diuretics promote the loss of water at the kidneys, lowering blood volume, and vasodilators further reduce blood pressure. A new class of antihypertensive drugs lowers blood pressure by preventing the conversion of angiotensin I to angiotensin II. These **angiotensin-converting enzyme (ACE) inhibitors**, such as *captopril*, are being used to treat chronic hypertension and congestive heart failure.

In **hypotension** blood pressure declines, and peripheral systems begin to suffer from oxygen and nutrient deprivation. One clinically important form of hypotension can develop following the administration of antihypertensive drugs. Problems may appear when the individual changes position, going from lying down to sitting, or sitting to standing. Normally each time you sit or stand, blood pressure in the carotid sinus drops, for the heart must suddenly counteract gravity to push blood up to the brain. The fall in pressure triggers the carotid reflex, and blood pressure returns to normal. But if the carotid response is prevented by beta-blockers or other drugs, blood pressure at the brain may fall so low that the individual becomes weak, dizzy, disoriented, or unconscious. This condition is known as **orthostatic hypotension** (ōr-tho-STAT-ik; *orthos*, straight + *statikos*, causing to stand), or simply **orthostasis** (ōr-thō-STĀ-sis). Most readers will have experienced brief episodes of orthostasis when standing up suddenly after reclining for an extended period.

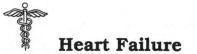

The initial symptoms of heart failure vary depending on whether the problem is restricted to the left ventricle or the right ventricle, or involves both. However, over time these differences are eliminated; for example, the major cause of right ventricular failure is left ventricular failure. Figure 21-B provides a simplified flow chart for heart failure and indicates potential therapies.

For example, suppose that the left ventricle cannot maintain normal cardiac output, due to damage to the ventricular muscle (see the discussion of myocardial infarctions on p. 646 of the text) or high arterial pressures (hypertension, discussed above). In effect, the left ventricle can no longer keep pace with the right ventricle, and blood backs up into the pulmonary circuit. This *venous congestion* is responsible for the term **congestive heart failure**. The right ventricle now works harder, elevating pulmonary arterial pressures and forcing blood through the lungs and into the weakened left ventricle.

At the capillaries of the lungs, arterial and venous pressures are now elevated. This elevated pressure shifts the dynamic center far toward the venous end of the capillaries, and fluid leaves the blood and enters the tissues of the lungs. The fluid buildup and compression of the airways reduces the effectiveness of gas exchange, leading to shortness of breath, often the first obvious sign of congestive failure. This fluid buildup begins at a postcapillary pressure of around 20 mm Hg.[1]

Over time, the less muscular right ventricle may become unable to generate enough pressure to force blood through the pulmonary circuit. Venous congestion now occurs in the systemic circuit, and cardiac output declines further. When the reduction in systemic pressures lowers blood flow to the kidneys, renin and erythropoietin are released. This in turn elevates blood volume, due to increased salt and water retention at the kidneys, and accelerated red blood cell production. This rise in blood volume actually complicates the situation, as it tends to increase venous congestion and cause widespread edema.

The increased volume of blood in the venous system leads to a distension of the veins, making superficial veins more prominent. When the heart contracts, the rise in pressure at the right atrium produces a pressure pulse in the large veins. This venous pulse can be seen and felt most easily over the right external jugular vein.

Treatment of congestive heart failure often includes:

---

[1] At a pressure of around 30 mm Hg fluid not only enters the tissues of the lungs, but crosses the alveolar walls and begins filling the airspaces. This condition is called **pulmonary edema**.

1. Restriction of salt intake. The expression "water follows salt" applies here, because when sodium and chloride ions are absorbed across the lining of the digestive tract, water is also absorbed by osmosis.

2. Administration of drugs to promote fluid loss. These drugs, called **diuretics** (dī-ū-RET-iks; *diouretikos*, promoting urine), increase salt and water losses at the kidneys. (The mechanism is described in Chapter 27 of the text.)

3. Extended bed rest, to enhance venous return to the heart, coupled with physical therapy to maintain good venous circulation.

FIGURE 21-B
Heart failure: symptoms and treatments

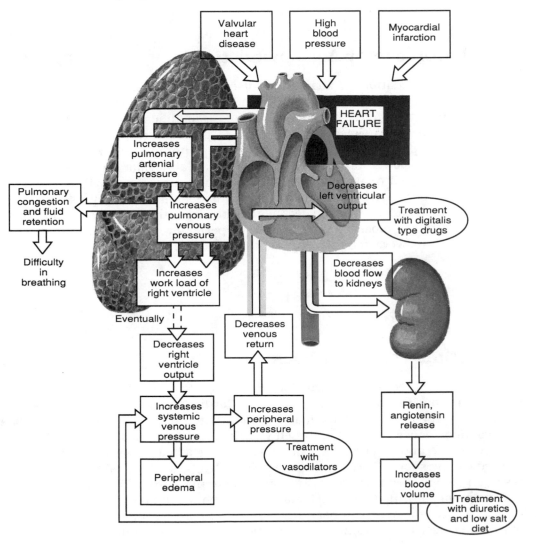

4. Administration of drugs that enhance cardiac output. These drugs may target the heart, the peripheral circulation, or some combination of the two. When the heart has been weakened, drugs related to *digitalis*, an extract from the leaves of the foxglove plant, are often selected. *Digitoxin*, *digoxin*, and *ouabain* are examples. These compounds increase the force of cardiac muscle fiber contractions. Drugs that cause venoconstriction may also be used, to enhance systemic venous return as diuretics lower the blood volume. When high blood pressure is a factor, some type of vasodilator is given.

5. Administration of drugs that reduce peripheral vascular resistance, such as hydralazine or ACE inhibitors. The drop in peripheral resistance reduces the workload of the left ventricle.

Baroreceptor Accommodation   *Page 698*

Receptor accommodation, or adaptation, discussed in Chapter 17 of the text (see pp. 526–527) occurs when a receptor becomes less sensitive to a chronic stimulus over time. Accommodation of baroreceptors is a significant complicating factor in the treatment of hypertension. When blood pressure rises or falls suddenly, the receptors respond by altering their rate of firing appropriately. But if the stimulus changes very slowly, or remains constant at an abnormal level, the receptors gradually adjust their response to accept the new pressure as "normal." The baroreceptors of a hypertensive individual usually consider the elevated pressures to be perfectly normal, and any sudden declines will be met with reflex opposition. Initial treatment with beta-blockers and other antihypertensive drugs must therefore be strong enough to overcome the baroreceptor response.

Other Types of Shock   *Page 704*

Although the text focuses on circulatory shock caused by low blood volume, shock can develop when the blood volume is normal. **Cardiogenic** (kar-dē-ō-JEN-ik) **shock** occurs when the heart becomes unable to maintain a normal cardiac output. The most common cause is failure of the left ventricle as a result of a myocardial infarction. Between 5 and 10 percent of patients surviving a heart attack must be treated for cardiogenic shock. The use of thrombolytic drugs, such as t-PA and streptokinase, can be very effective in restoring coronary circulation and relieving the peripheral symptoms. Cardiogenic shock may also be the result of arrhythmias, valvular heart disease, advanced coronary artery disease, cardiomyopathy, or ventricular arrhythmias (see Chapters 20 and 21 of the text and this Manual).

In **obstructive shock**, ventricular output is reduced because tissues or fluids are restricting the expansion and contraction of the heart. For example, fluid buildup in the pericardial cavity (*cardiac tamponade*, discussed in Chapter 20 of this Manual (see Infection and Inflammation of the Heart), can compress the heart and limit ventricular filling.

Distributive shock results from a widespread, uncontrolled vasodilation. This produces a dramatic fall in blood pressure that leads to a reduction in blood flow and the onset of shock. Three important examples are *neurogenic shock*, *septic shock*, and *anaphylactic shock*.

Neurogenic (noo-rō-JEN-ik) **shock** can be caused by general or spinal anesthesia and by trauma or inflammation of the brain stem. The underlying problem is damage to the vasomotor center or to the sympathetic tracts or nerves, leading to a loss of vasomotor tone.

Septic shock results from the massive release of endotoxins, poisons derived from the cell walls of bacteria during a systemic infection. These compounds cause a vasodilation of precapillary sphincters throughout the body, resulting in a drop in peripheral resistance and a decline in blood pressure. Symptoms of septic shock generally resemble those of other types of shock, but the skin is flushed, and the individual has a high fever. For this reason septic shock is also known as "warm shock." One interesting example of septic shock, called **toxic shock syndrome** (TSS) results from an infection by the bacterium *Staphylococcus aureus*. This disease was unrecognized before 1978, when it appeared in a group of children. Since that time there have been roughly 2500 cases in the United States, most (95 percent) affecting women. Although other sources of infection are possible, infection most often appears to occur during menstruation, and the chances of infection are increased with the use of superabsorbent tampons. (The brands involved have been taken off the market, and the incidence has declined steadily since 1980.)

Extensive peripheral vasodilation also occurs in **anaphylactic** (an-a-fi-LAK-tik) **shock**, a dangerous allergic reaction. This type of shock is discussed in Chapter 22 of the text (see p. 744).

The Lymphatic System and Immunity

Lymphedema
Page 712

Blockage of the lymphatic drainage from a limb produces **lymphedema** (lim-fe-DĒ-ma). In this condition, interstitial fluids accumulate, and the limb gradually becomes swollen and grossly distended. If the condition persists the connective tissues lose their elasticity, and the swelling becomes permanent. Lymphedema is painless, and the condition by itself does not pose a major threat to life. The danger comes from the continual risk of an uncontrolled infection developing in the affected area. Because the interstitial fluids are essentially stagnant, toxins and pathogens can accumulate and overwhelm the local defenses without fully activating the immune system in the rest of the body.

Temporary lymphedema of the feet and ankles may be caused by tight clothing constricting the lymphatic circulation or by prolonged standing or sitting. Elevating the feet and loosening clothing may eliminate the problem. Chronic lymphedema usually results as scar tissue forms in an area of injury. Trauma, infections, and surgical procedures are often implicated. In **filariasis** (fil-a-RĪ-a-sis) a parasitic nematode (roundworm) carried by mosquitos forms massive colonies within lymphatic channels. Repeated scarring of the passageways eventually blocks lymphatic drainage and produces the extreme lymphedema with permanent distension of tissues known as **elephantiasis** (el-e-fan-TĪ-a-sis).

Therapy for chronic lymphedema consists of treating infections by the administration of antibiotics, and (when possible) reducing the swelling. One possible treatment involves the application of elastic wrappings that squeeze the tissue. This external compression elevates the hydrostatic pressure of the interstitial fluids and opposes the entry of additional fluid from the capillaries.

Infected Lymphatic Nodules
Page 715

The lymphocytes in a lymph nodule are not always able to destroy bacterial or viral invaders that have crossed the adjacent epithelium. When such pathogens become established in a lymph nodule, an infection develops. **Tonsillitis** is an infection of one of the tonsils, most often the pharyngeal tonsil. An individual with tonsillitis develops a high fever and leukocytosis (an abnormally high white blood cell count). The affected tonsil becomes swollen and inflamed, sometimes enlarging enough to partially block the entrance to the trachea. Breathing then becomes difficult, and in severe cases impossible. As the infection proceeds, abscesses develop within the tonsilar tissues, and the bacteria may enter the bloodstream by passing through the lymphatic capillaries and vessels to the venous system.

In the early stages, antibiotics may control the infection, but once abscesses have formed the best treatment involves surgical drainage of the abscesses. **Tonsillectomy**, the removal of the tonsil, was once highly recommended and frequently performed to prevent recurring tonsilar infections. The procedure does reduce the incidence and severity of subsequent infections, but questions have been raised concerning the overall cost to the individual. The tonsils represent a "first line" of defense against bacterial invasion of the pharyngeal walls. If they are removed, bacteria may not be detected until a truly severe infection is well under way.

Appendicitis usually follows an erosion of the epithelial lining of the appendix. Several factors may be responsible for the initial ulceration, notably bacterial or viral pathogens. Bacteria that normally inhabit the lumen of the large intestine then cross the epithelium and enter the underlying tissues. Inflammation occurs, and the opening between the appendix and the rest of the intestinal tract may become constricted. Mucus secretion accelerates, and the organ becomes increasingly distended. Eventually the swollen and inflamed appendix may rupture or perforate. If this occurs, the bacteria will be released into the warm, dark, moist confines of the abdominopelvic cavity, where they can cause a life-threatening infection. The most effective treatment for appendicitis is the surgical removal of the organ, a procedure known as an **appendectomy**.

Lymphomas *Page 715*

Lymphomas are malignant cancers consisting of abnormal lymphocytes or lymphocytic stem cells. Over 30,000 cases of lymphoma are diagnosed in the United States each year, and that number has been steadily increasing. There are many different types of lymphoma. One form, called **Hodgkin's disease** (**HD**), accounts for roughly 40 percent of all lymphoma cases. Hodgkin's disease most often strikes individuals at ages 15–35 or those over age 50. The reason for this pattern of incidence is unknown; although the cause of the disease is uncertain, an infectious agent (probably a virus) is suspected. Other types are usually lumped together under the heading of **non-Hodgkin's lymphoma** (**NHL**). They are extremely diverse, and in most cases the primary cause remains a mystery. At least some forms reflect a combination of inherited and environmental factors. For example, one form, called **Burkitt's lymphoma**, most often affects male children in Africa and New Guinea. The affected children have been infected with the *Epstein-Barr virus* (EBV).[1] The EBV infects B cells, but under normal circumstances the infected cells are destroyed by the immune system. EBV is widespread in the environment, and childhood exposure usually produces lasting immunity. Children developing Burkitt's lymphoma may have a genetic susceptibility to EBV infection; in addition, presence of another illness, such as malaria, may weaken their immune systems to the point that a lymphoma can develop.

The first symptom associated with any lymphoma is usually a painless enlargement of lymph nodes. The involved nodes have a firm, rubbery texture. Because the nodes are painless, the condition is often overlooked until it has progressed to the point that secondary symptoms appear. For example, patients seeking help for recurrent fever, night sweats, gastrointestinal or respiratory problems, or weight loss may be unaware of any underlying lymph node changes. In the late stages of the disease, symptoms can include liver or spleen enlargement, CNS dysfunction, pneumonia, a variety of skin conditions, and anemia.

In planning treatment, clinicians consider the histological structure of the nodes and the *stage* of the disease. When examining a biopsy, the structure of the node is described as *nodular* or *diffuse*. A nodular node retains a semblance of normal structure, with follicles and germinal centers. In a diffuse node the interior of the node has changed, and follicular structure has broken down. In general, the nodular lymphomas progress more slowly than the diffuse forms, which tend to be more aggressive. On the other hand, the nodular lymphomas are more difficult to treat and are more likely to recur even after remission has been achieved.

The most important factor influencing treatment selection is the stage of the disease. Table 22-A includes a simplified staging classification for lymphomas. When diagnosed early (stage I or II), localized therapies may be effective. For example, the cancerous node(s) may be surgically removed and the region(s) irradiated to kill residual cancer cells. Success rates are very high when a lymphoma is detected in these early stages. For Hodgkin's disease, localized radiation can produce remission lasting 10 years or more in over 90 percent of patients. Treatment of localized NHL is somewhat less effective. The 5-year remission rates average 60–80 percent for all types; success rates are higher in nodular forms than for diffuse forms.

Although these are encouraging results, it should be noted that few lymphoma patients are diagnosed while in the early stages of the disease. For example, only 10–15 percent of NHL patients are diagnosed at stages I or II. For lymphomas at stages III and IV, treatment most often involves chemotherapy. Combination che-

---

[1] This highly variable virus is also responsible for *infectious mononucleosis*, (discussed further below), and it has been suggested as the primary cause of *chronic fatigue syndrome*.

Stage I: Involvement of a single node or region (or of a single extranodal site)
   Typical treatment: surgical removal and/or localized irradiation; in slowly progressing forms of NHL, treatment may be postponed indefinitely.

Stage II: Involvement of nodes in two or more regions (or of an extranodal site and nodes in one or more regions) on the same side of the diaphragm
   Typical treatment: surgical removal and localized irradiation that includes an extended area around the cancer site (the *extended field*).

Stage III: Involvement of lymph node regions on both sides of the diaphragm. This is a large category that is subdivided on the basis of the organs or regions involved. For example, in stage III$_s$ the spleen contains cancer cells.
   Typical treatment: combination chemotherapy, with or without radiation; radiation treatment may involve irradiating all of the thoracic and abdominal nodes plus the spleen (*total axial nodal irradiation*, or *TANI*).

Stage IV: Widespread involvement of extranodal tissues above and below the diaphragm
   Treatment is highly variable, depending on the circumstances. Combination chemotherapy is always used; it may be combined with whole-body irradiation. The "last resort" treatment involves massive chemotherapy followed by a bone marrow transplant.

motherapy, in which two or more drugs are administered simultaneously, is the most effective treatment. For Hodgkin's disease, a four-drug combination with the acronym MOPP (*nitrogen **m**ustard*, ***O**ncovin* [*vincristine*], ***p**rednisone*, and ***p**rocarbazine*) produces lasting remission in 80 percent of patients.

Bone marrow transplantation is a treatment option for acute, late-stage lymphoma. When suitable donor marrow is available, the patient receives whole-body irradiation, chemotherapy, or some combination of the two sufficient to kill tumor cells throughout the body. This treatment also destroys normal bone marrow cells. Donor bone marrow is then infused, and over the next 2 weeks the donor cells colonize the bone marrow and begin producing red blood cells, granulocytes, monocytes, and lymphocytes.

Potential complications of this treatment include the risk of infection and bleeding while the donor marrow is becoming established. The immune cells of the donor marrow may also attack the tissues of the recipient, a response called *graft versus host disease*, or *GVH*. For a patient with stage I or II lymphomas, without bone marrow involvement, bone marrow can be removed and stored (frozen) for over 10 years. If other treatment options fail, or the patient comes out of remission at a later date, an autologous marrow transplant can be performed. This eliminates the need for donor typing and the risk of GVH disease.

Disorders of the Spleen

Page 720

An impact to the left side of the abdomen can distort or damage the spleen. Such injuries are known risks of contact sports, such as football or hockey, and more solitary athletic activities, such as skiing or sledding. However, the spleen tears so easily that a seemingly minor blow to the side may rupture the capsule. The result is serious internal bleeding and eventual circulatory shock.

Because the spleen is relatively fragile, it is very difficult to repair surgically. (Sutures usually tear out before they have been tensed enough to stop the bleeding.) Treatment for a severely ruptured spleen involves its complete removal, a process called a **splenectomy** (sple-NEK-to-mē). Removal of the spleen in an adult usually has little effect on immune defenses because virtually all of the spleen's functions are also performed by other lymphatic organs.

The spleen responds like a lymph node to infection, inflammation, or invasion by cancer cells. The enlargement that follows is called **splenomegaly** (splen-ō-MEG-a-lē; *megas*, large), and splenic rupture may also occur under these conditions. One relatively common condition causing splenomegaly is **mononucleosis**. This condition, also known as the "kissing disease," results from chronic infection by the *Epstein-Barr virus* (*EBV*). In addition to splenic enlargement, symptoms of mononucleosis include fever, sore throat, widespread swelling of lymph nodes, increased numbers of lymphocytes in the blood, and the presence of circulating antibodies to the virus. The condition most often affects young adults (age 15–25) in the spring or fall. Treatment is symptomatic, as there are no drugs that are effective against this virus. The most dangerous aspect of the disease is the risk of rupturing the enlarged spleen, which becomes fragile. Patients are therefore cautioned against heavy exercise or other activities that increase abdominal pressures. If the spleen does rupture, severe hemorrhaging may occur; death will follow unless transfusion and an immediate splenectomy can be performed.

An individual whose spleen is missing or nonfunctional has **hyposplenism** (hī-pō-SPLĒN-ism). Hyposplenism usually does not pose a serious problem, but such individuals are more prone to certain bacterial infections, and special immunizations may be recommended. In **hypersplenism** the spleen becomes overactive, and the increased phagocytic activities lead to anemia (low number of RBCs), leukopenia (low numbers of WBCs), and thrombocytopenia (low numbers of platelets). Splenectomy is the only known cure for hypersplenism.

AIDS

Page 732

Acquired immune deficiency syndrome, or **AIDS**, develops following infection by a virus known as **human immunodeficiency virus**, or **HIV**. There are at least three different types of AIDS virus, designated HIV-1, HIV-2, and HIV-3. Most AIDS patients in the United States are infected with HIV-1.[1] The discussion that follows, based on information pertaining to HIV-1 infection, expands upon the discussion on p. 732 of the text.

Symptoms of the disease   The initial infection usually triggers production of antibodies against the virus. These antibodies appear in the serum within 2–6 months of exposure, and antibody tests can be used to detect infection (see below). Further symptoms may not appear for 5–10 years or more. When symptoms do appear they are often mild, consisting of lymphadenopathy, chronic nonfatal infections, diarrhea, and weight loss. This condition has been called **AIDS-related complex**, or **ARC**. It is not known what triggers the conversion from the carrier state to ARC, and from ARC to AIDS.

HIV-1 selectively infects helper T cells. This impairs the immune response, and the effect is magnified because suppressor T cells are relatively unaffected by the virus. Over time, circulating antibody levels decline, cellular immunity is reduced, and the body is left without defenses against a wide variety of bacterial, viral, fungal, and protozoan invaders.

This vulnerability is what makes AIDS dangerous—the effects of HIV on the immune system are not *by themselves* life-threatening, but the infections that result certainly are. With the depression of immune function, ordinarily harmless pathogens can initiate lethal infections, known as *opportunistic infections*. In fact, the most common and dangerous pathogens for an AIDS patient are microorganisms that seldom cause illnesses in humans with normal immune systems. AIDS patients are especially prone to lung infections and pneumonia, often caused by infection with the fungus *Pneumocystis carinii*. They are also subject to a variety of other fungal infections, such as cryptococcal meningitis, and an equally broad array of bacterial and viral infections. The symptoms and time course of these infections are very different from those in normal individuals, because AIDS patients are virtually defenseless.

In addition to problems with pathogenic invasion, immune surveillance is depressed, and the risk of cancer increases. One of the most common cancers seen in AIDS patients is **Kaposi's sarcoma**, a condition that is extremely rare in uninfected individuals. Kaposi's sarcoma typically begins with rapid cell divisions in endothelial cells of cutaneous blood vessels. Associated lesions, blue or a deep brown-purple in color, generally appear first in the hands or feet, and later occur closer to the trunk. In a small number of patients the lesions develop in the epithelium of the digestive or respiratory tracts. In normal individuals the tumor usually

---

[1] HIV-2 infections are most common in Africa. Because individuals infected with HIV-2 may not always develop AIDS, it is thought that this may be a less dangerous virus (so far 100 percent of those infected with HIV-1 eventually develop AIDS). The distribution and significance of HIV-3 infection remain to be determined.

does not metastasize; in the case of AIDS patients, whose immune systems are relatively ineffective, it often converts to an aggressive, invasive cancer.

If the AIDS patient survives all of these assaults, infection of the CNS by HIV eventually produces neurological disorders and a progressive dementia. So far, AIDS is invariably fatal, and it appears likely that all those who carry the virus will eventually die of the disease.

Incidence   At the end of 1990 there were over 152,000 known cases of AIDS in the United States, almost four times the number reported just 3 years earlier, and the number of deaths had exceeded 100,000. The Centers for Disease Control (CDC) in Atlanta, which has been monitoring the spread of AIDS, estimates that the number of AIDS cases in the United States will continue to increase rapidly, reaching 480,000 cases by the end of 1993 (Figure 22-A).

Because the virus can remain in the body for years without producing clinical symptoms, the number of individuals infected and at risk is certain to be far higher. Estimates of the number of infected individuals in the U.S. population range from 1 to 2 million; barring a major clinical advance (see below) most or all of these individuals will develop ARC and AIDS within the next decade. The numbers worldwide are even more frightening. The World Health Organization estimates that as many as 10 million people may be infected, and the number of AIDS patients worldwide continues to climb rapidly. Several African nations are already on the verge of social and economic collapse because of devastation by AIDS. In Malawi, one-third of the population is infected, and 60 percent of pregnant women carry the virus.

Modes of infection   Infection with HIV occurs through intimate contact with the body fluids of infected individuals. Although all body fluids, including saliva, tears, and breast milk, carry the virus, the major routes of infection involve contact with blood, semen, or vaginal secretions. The transmission pattern has been analyzed for U.S. adult and adolescent AIDS patients. Four major transmission routes have been identified, and we will consider them individually.

Sexual transmission   In approximately 62 percent of all cases, exposure occurred through sexual contact with infected individuals. Male homosexual contact was most often involved (57 percent versus 5 percent for heterosexual contact), and the ratio of male to female AIDS patients is approximately 9 to 1. A comparable transmission pattern is found in Canada, Europe, South America, Australia, and New Zealand. In the Caribbean and in Africa, by contrast, AIDS began in the heterosexual community. It affects heterosexual men and women in roughly equal numbers, and the sex ratio for AIDS patients in these countries is approximately 1 to 1.

Two factors may account for the pattern of transmission observed in the United States. First, the disease appears to have spread through the homosexual commu-

84

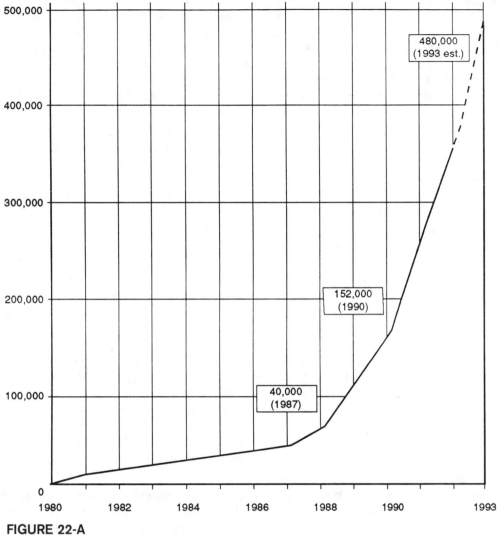

FIGURE 22-A
AIDS Cases in the United States

nity before entering the heterosexual population via bisexual males. Second, there is statistical evidence that on a per-exposure basis the risk of male-to-male or male-to-female transmission is greater than the risk of female-to-male transmission. As a result, it may take longer to spread the virus through the heterosexual population (male 1 to female 1 to male 2) than through the homosexual population (male 1 to male 2 to male 3). Whether homosexual or heterosexual contact is involved, a sex partner whose epithelial defenses are weakened is at increased risk of infection. This accounts for the relatively higher rate of transmission through anal intercourse, which often damages the delicate lining of the anorectal canal.

Because of the predominance of homosexual transmission in this country, many people still consider AIDS to be a homosexual disease. It is not. Over time, the number of cases in the heterosexual population has been steadily increasing; since 1987, the percentage of homosexual or bisexual AIDS patients has dropped by roughly 25 percent, and the number of cases transmitted by heterosexual contact has steadily increased. It can be anticipated that over time the sex ratio will continue to shift toward the 1:1 male-to-female ratio typical of Africa and the Caribbean.

Intravenous drug use   Roughly 19 percent of AIDS patients contracted the disease through the shared use of needles. (Another 6 percent were homosexual males who shared needles with other drug users, making it unclear which factor was responsible for transmission.) Although only small quantities of blood are inadvertently transferred when needles are shared, this practice injects the AIDS virus directly into the bloodstream. It is thus a very effective way to transmit the disease.

Receipt of blood or tissue products   About 3 percent of AIDS patients became infected with the virus after receiving a transfusion of contaminated whole blood or plasma, an infusion of blood products, such as platelets or extracts of pooled sera, or an organ transplant from an infected individual. With careful screening of blood and blood products, the rate of transmission via this route is declining.

Prenatal exposure   Approximately 2000 infants are born each year already infected with AIDS. Although this is a relatively small number, compared with the total number of AIDS patients in the United States, it is increasing rapidly. A pregnant woman infected with the AIDS virus has a 20–33 percent chance of infecting her baby across the placenta or at delivery. As AIDS spreads through the heterosexual population, more pregnant women will become infected, and maternal-fetal transmission will become more common. These unfortunate infants will place social and financial stresses on our society for the rest of the 1990s.

Prevention of AIDS   The best defense against AIDS consists of avoiding exposure to the virus. The most important rule is to *avoid sexual contact with infected individuals.* All forms of sexual intercourse carry the potential risk of viral transmission. The use of synthetic (latex) condoms has been recommended when the previous history of a sex partner is not known. (Condoms that are not made of synthetic materials are effective in preventing pregnancy but do not block the passage of viruses.) Although condom use does not provide absolute protection, it drastically reduces the risk of infection.

Attempts are under way to ensure that blood and blood products are adequately screened for the presence of HIV-1. A simple blood test exists for the detection of HIV-1 antibodies, and a positive reaction indicates previous exposure to the virus. The assay, an example of an **ELISA test** (enzyme-linked immunoabsorbent assay), is now used to screen blood donors, reducing the risk of infection by transfusion or the use of blood products from pooled sera. Pooled sera can also be heat-treated by exposing it to temperatures sufficient to kill the virus but too low to permanently denature blood proteins.

Most public health facilities will perform the test on request for individuals who fear that they may have already been exposed to the AIDS virus. Unfortunately the test is not 100 percent reliable, and false positive reactions occur at a rate of about 0.4 percent. In addition, the ELISA test does not detect HIV-2 or HIV-3. In the event of a positive test result, a retest should be performed using the more sensitive *western blot* procedure. It should be recognized that because of the variable incubation period, a positive test for HIV infection does not mean that the individual has AIDS, or even ARC. It does mean that the individual is likely to develop these conditions at some time in the future, and that *the person is now a carrier and capable of infecting others.* By the time an individual develops AIDS, he or she is obviously sick and usually has little interest in sexual activity. In terms of the spread of this disease, the most dangerous individuals are those who appear perfectly healthy and have no idea that they are carrying the virus.

Despite intensive efforts, a vaccine has yet to be developed that will provide immunity from HIV infection. Current research programs are attempting to stimulate antibody production in response to (1) killed but intact viruses, (2) fragments of the viral envelope, (3) HIV proteins on the surfaces of other, less dangerous viruses, or (4) T cell proteins that are targeted by HIV. (The last approach is based on the hypothesis that the antibodies produced will cover the binding sites, preventing viral attachment and penetration.) As yet none of these approaches has been successful enough to warrant widespread clinical trials.

Treatment There is no cure for ARC or AIDS. However, the length of survival for AIDS patients has been steadily increasing because (1) new drugs are available that slow the progress of the disease and (2) improved antibiotic therapies have helped overcome infections that would otherwise prove fatal. This combination is extending the lifespan of patients while the search for more effective treatment continues. (It should be noted, however, that overcoming an infection in an AIDS patient with antibiotics may require doses up to 10 times greater than those used to fight infections in normal individuals. Moreover, once the infection has been overcome, *the patient may have to continue taking that drug for the rest of his or her life.* As a result, some AIDS patients find themselves taking 50–100 pills per day just to prevent recurrent infections.

An antiviral drug, **azidothymidine** or **AZT** (*Zidovudine* or *Retrovir*), can decrease the symptoms of ARC and slow the progression of AIDS. Low doses of AZT have now been shown to be effective in treating patients with ARC. Higher doses are used to treat AIDS, but this use can lead to a variety of unpleasant side effects, including anemia and even bone marrow failure. However, AZT is effective in reducing the neurological symptoms of AIDS, because it can cross the blood-brain barrier to reach infected CNS tissues. By alternating AZT treatment with another antiviral drug, **dideoxycytidine** (**ddC**), side effects of both drugs are reduced. Other drugs are undergoing clinical testing in the United States and Europe. But, although the next generation of drugs will further extend the lives of ARC and AIDS patients, none of the drugs under review promises a cure for the disease.

AIDS and the cost to society   Treatment of AIDS is complex and expensive. The average cost is $70,000 per patient per year, giving a projected annual cost of roughly $34 billion by 1993. As more individuals become infected, and more of the 1–2 million people already infected develop ARC or AIDS, the costs will continue to escalate rapidly.

There are other, less obvious costs to society. Individuals with depressed immune systems get sick more often and stay sick longer. This makes them more likely to spread the infecting pathogen to other, healthy individuals. Some diseases, such as tuberculosis, that were previously present at very low levels in the U.S. population, are now occurring with increased frequency. In part this is because AIDS patients are succumbing to infection, but it also reflects the spread of these diseases via transmission from AIDS patients into the general population.

Lyme Disease   *Page 738*

In November 1975, the town of Lyme, Connecticut experienced an epidemic of adult and juvenile arthritis. Between June and September, 59 cases were reported, 100 times the statistical average for a town of that size. Symptoms were unusually severe; in addition to joint degeneration, victims experienced chronic fever and a prominent rash that began as a bull's-eye centered around what appeared to be an insect bite (Figure 22-B). This combination was enough to justify a full-scale investigation that continued for almost 2 years before the pathogen and transmission route were tracked down. Lyme disease is caused by a bacterium, *Borrelia burgdorferi*, transmitted through the bite of a deer tick. The high rate of infection among children reflects the fact that they play outdoors during their summer vacations. After 1975, the problem became regional and then national in scope. In 1990, there were some 30,000 people with Lyme disease in the United States alone. Most cases are reported from the Northeast coast, the Midwest, and the West Coast of the United States.

Although some of the joint destruction may result from immune complex deposition, in a mechanism comparable to that involved in rheumatoid arthritis, many of the symptoms (fever, pain, skin rash) develop in response to the release of interleukin-1 (Il-1) by activated macrophages. The cell walls of *B. burgdorferi* contain lipid-carbohydrate complexes that stimulate secretion of Il-1 in large quantities. By stimulating the specific and nonspecific defense mechanisms of the body, Il-1 exaggerates the inflammation, rash, fever, pain, and joint degeneration associated with

88

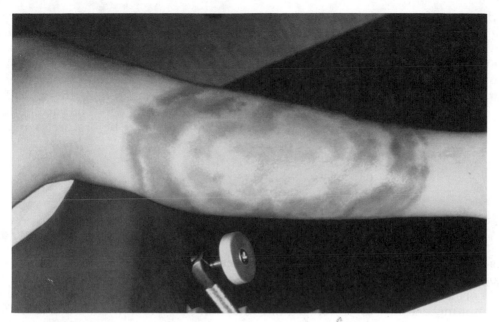

FIGURE 22-B
Lyme disease rash

the primary infection. Treatment for Lyme disease consists of administering antibiotics and anti-inflammatory drugs.

Fetal Infections *Page 740*

Fetal infections are rare because the developing fetus has passive immunity courtesy of the antibodies produced by the mother. These defenses break down if the maternal antibodies are unable to cope with a bacterial or viral infection. For example, the microorganisms responsible for *syphilis* and *rubella* ("German measles") can cross from the maternal to the fetal bloodstream, producing a congenital infection that leads to the production of fetal antibodies. When this occurs after the fourth developmental month, the fetus responds by producing IgM antibodies; the ability to produce IgG in response to pathogenic infection does not develop until several months after birth.

Blood drawn from a newborn infant or taken from the umbilical cord of a developing fetus can be tested for the presence of IgM antibodies. This procedure provides concrete evidence of congenital infection. For example, a newborn infant with congenital syphilis will have IgM antibodies that target the pathogenic bacterium involved (*Treponema pallidum*). Fetal or neonatal (newborn) blood may also be tested for antibodies against the rubella virus or other pathogens.

In the case of congenital syphilis, antibiotic treatment of the mother can prevent fetal damage. In the absence of antibiotic treatment, fetal syphilis can cause liver and bone damage, hemorrhaging, and a susceptibility to secondary infections in the newborn infant. There is no satisfactory treatment for congenital rubella infection, which can cause severe developmental abnormalities. For this reason rubella immunization has been recommended for young children (to slow the spread of the disease in the population) and for women of childbearing age. The vaccination, which contains live attenuated virus, must be administered *before* pregnancy to prevent maternal infection during pregnancy.

SCID   *Page 741*

In **severe combined immunodeficiency disease** (**SCID**) the individual fails to develop either cellular or humoral immunity. Lymphocyte populations are reduced, and normal B and T cells are not present. Patients with SCID are unable to provide an immune defense, and even a mild infection can prove fatal. Total isolation offers protection at great cost and with severe restrictions on the individual's lifestyle. Bone marrow transplants from compatible donors, usually a close relative, have been used to colonize lymphatic tissues with functional lymphocytes. (An experimental approach involving gene-splicing techniques was described on p. 117 of the text.)

The most famous SCID patient was the "bubble boy," David, shown in Figure 22-C. He was kept in physical isolation until age 12, when he received a bone marrow transplant. Before the donor marrow cells established a functional immune system, David died of cancer. Although technology had protected him from external pathogens, without immune surveillance he had no defense against threats from within.

Systemic Lupus Erythematosus   *Page 742*

Systemic lupus erythematosus (LOO-pus e-rith-ē-ma-TŌ-sis), or **SLE**, appears to result from a generalized breakdown in the antigen recognition mechanism. An individual with SLE manufactures autoantibodies against nucleic acids, ribosomes, clotting factors, blood cells, platelets, and lymphocytes. The immune complexes form deposits in peripheral tissues, producing anemia, kidney damage, arthritis, and vascular inflammation. CNS function deteriorates if the blood flow through the damaged vessels slows or stops.

The most obvious sign of this condition is the presence of a butterfly-shaped discoloration of the face, centered over the bridge of the nose (Figure 22-D). SLE

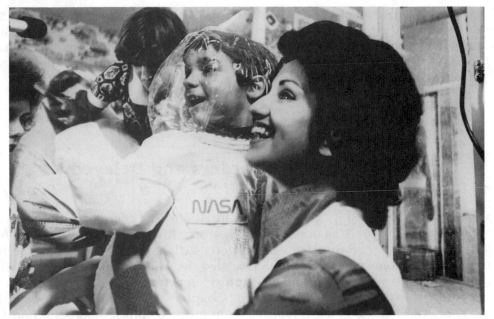

FIGURE 22-C
David, the "boy in the bubble"

FIGURE 22-D
Rash of systemic lupus erythematosus

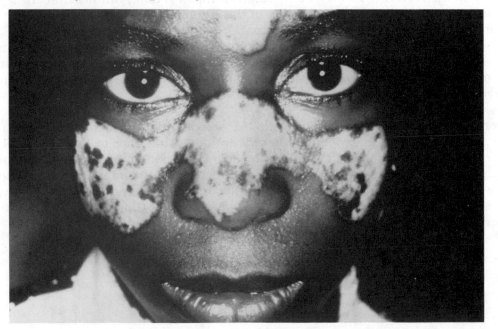

affects women nine times as often as men, and the U.S. incidence averages 2–3 cases per 100,000 population. There is no known cure, but almost 80 percent of SLE patients survive 5 years or more after diagnosis. Treatment consists of controlling the symptoms and depressing the immune response through administration of specialized drugs or corticosteroids.

Immune Complex Disorders and Delayed Hypersensitivity   *Page 742*

In Chapter 22 of the text (p. 742) it was pointed out that there are four classes of allergic reactions. Type I, or immediate hypersensitivity, discussed on pp. 744–745, is responsible for relatively mild conditions such as hay fever and more serious ones such as anaphylaxis and asthma (see also the Clinical Comment on p. 768). An important example of type II, or cytotoxic reactions—the cross-reactions of incompatible blood types—is discussed in Chapter 19 of the text (see pp. 617–619). The other two classes are type III, or immune complex disorders, and type IV, or delayed hypersensitivity.

Under normal circumstances, immune complexes are promptly eliminated by phagocytosis. But when an antigen appears suddenly, in high concentrations, the local phagocytic population may not be able to cope with the situation. The immune complex then enlarges further, eventually forming insoluble granules that are deposited in the affected area. The presence of these complexes triggers extensive activation of complement, leading to inflammation and tissue damage at that site. This process is further enhanced by neutrophils, which release enzymes that attack the cells and tissues adjacent to the immune complex. The most serious **immune complex disorders** involve deposits within blood vessels and in the filtration membranes of the kidney.

Delayed hypersensitivity begins with the sensitization of cytotoxic T cells. AT the initial exposure, macrophages present antigenic materials to T cells. On subsequent exposure, the T cells respond by releasing lumphokines. These lymphokines stimulate macrophage activity and produce a massive inflammatory response in the immediate area. Examples of delayed hypersensitivity include the many types of contact dermatitis, such as poison ivy, discussed in Chapter 6 of this Manual (see Dermatitis).

CHAPTER 23

The Respiratory System

Nosebleeds

Page 751

The extensive vascularization of the nasal cavity and the relatively vulnerable position of the nose make a nosebleed, or **epistaxis** (ep-i-STAK-sis), relatively common. Bleeding usually involves vessels of the mucosa covering the cartilaginous portion of the septum. Packing the vestibule with gauze or pinching the external nares together to squeeze the vessels against the septum will often control the bleeding until clotting occurs. More severe bleeding originating elsewhere in the nasal cavity may require packing the posterior portion of the cavity via the internal nares.

Epistaxis can result from any factor affecting the integrity of the epithelium or the underlying vessels. Examples would include trauma, such as a punch in the nose, drying, infections, allergies, or clotting disorders. Hypertension may also provoke a nosebleed by rupturing small vessels of the lamina propria.

Disorders of the Larynx

Page 753

Infection or inflammation of the larynx is known as **laryngitis** (lar-in-JĪ-tis). This condition often affects the vibrational qualities of the vocal cords; hoarseness is the most familiar symptom. Mild cases are temporary and seldom serious, but bacterial or viral infection of the epiglottis or upper trachea in children can be very dangerous because the swelling may close the glottis and cause suffocation. Acute **epiglottitis** (ep-i-glot-TĪ-tis) can develop relatively rapidly following a bacterial infection of the throat. Although most common in children, it does occur in adults, especially those with *Hodgkin's disease* or *leukemia*, two cancers described in Chapter 22 of this Manual. Serious inflammation and edema of the trachea can also occur in small children following infection with one of the *parainfluenza viruses*. The condition, *laryngotracheobronchitis*, is more commonly called **croup** (kroop).

93

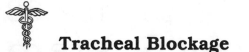

Tracheal Blockage Page 754

Foreign objects that become lodged in the larynx or trachea are usually expelled by coughing. If the individual can speak or make a sound, the airway is still open, and no emergency measures should be taken. If the victim can neither breathe nor speak, an immediate threat to life exists. Unfortunately many victims become acutely embarrassed by this situation, and rather than seek assistance, they run to the nearest restroom and quietly expire.

In the **Heimlich** (HĪM-lik) **maneuver**, or *abdominal thrust*, a rescuer applies compression to the abdomen just beneath the diaphragm. This elevates the diaphragm forcefully and may generate enough pressure to blow the offending object out of the trachea and glottis. This maneuver must be performed properly to avoid damage to internal organs. Organizations such as the Red Cross, the local fire department, or other charitable groups usually hold brief training sessions throughout the year.

If the blockage remains, professionally qualified rescuers may perform a **tracheostomy** (trā-kē-OS-to-mē). In this procedure an incision is made through the anterior tracheal wall, and a tube is inserted. The tube bypasses the larynx and permits air to flow directly into the trachea. A tracheostomy may also be required when the larynx becomes blocked by a foreign object, inflammation, or sustained laryngeal spasms, or when a portion of the trachea has been crushed.

Examining the Living Lung Page 756

A chest X-ray remains the standard diagnostic screening test. This procedure can detect abnormalities in lung structure including scar tissue formation, fluid accumulation, or distortion of the conducting passageways. For example, the fluid buildup that can occur during congestive heart failure (see Chapter 21 of this Manual) or a pleural effusion (p. 155 of the text) can be detected on a standard X-ray. CT scans show much greater definition of internal structures and can clarify the nature of abnormalities initially spotted on a chest X-ray. CT scans are particularly helpful in diagnosing and tracking the progression of lung cancers. **Lung scans** are made following the injection of radioactive tracers or inhalation of radioactive gases. These procedures (discussed in Chapter 2 of the text) can detect abnormalities in air flow or pulmonary circulation.

Bronchoscopy (brong-KOS-ko-pē) involves the insertion of a fiberoptic bundle, or **bronchoscope**, a few millimeters in diameter into the trachea. The bundle, once inserted, can be steered along the conducting passageways to the level of the smaller bronchi. In addition to permitting direct visualization of bronchial structures, the bronchoscope can collect tissue or mucus samples from the respiratory tract. In **bronchography** (brong-KOG-ra-fē) a bronchoscope or catheter introduces a radiopaque material into the bronchi. This technique can permit detailed analysis of bronchial masses, such as tumors, or other obstructions along the bronchial tree.

Pneumonia   *Page 756*

Pneumonia (nū-MŌ-nē-a) develops from a pathogenic infection of the lobules of the lung. As inflammation occurs within the lobules, respiratory function deteriorates as a result of fluid leakage into the alveoli and/or swelling and constriction of the respiratory bronchioles. When bacteria are involved, they are usually species normally found in the mouth and pharynx that have somehow managed to evade the respiratory defenses. As a result, pneumonia becomes more likely when the respiratory defenses have been compromised by other factors, such as epithelial damage from smoking or the breakdown of the immune system in AIDS. The most common pneumonia that develops in AIDS patients results from infection by the fungus *Pneumocystis carinii*. These organisms are normally found in the alveoli, but in healthy individuals the respiratory defenses are able to prevent infection and tissue damage.

Tuberculosis   *Page 758*

Tuberculosis (tū-ber-kū-LŌ-sis), or **TB**, results from a bacterial infection of the lungs, although other organs may be invaded as well. The bacteria, *Mycobacterium tuberculosis*, may colonize the respiratory passageways, the interstitial spaces, and/or the alveoli. Symptoms are variable, but usually include coughing and chest pain, with fever, night sweats, fatigue, and weight loss.

At the site of infection, macrophages and fibroblasts proceed to wall off the area, forming an abscess. If the scar-tissue barricade fails, the bacteria move into the surrounding tissues and the process repeats itself. The resulting masses of fibrous tissue distort the conducting passageways, increasing resistance and decreasing airflow. In the alveoli, the attacked surfaces are destroyed. The combination reduces the vital capacity and the area available for gas exchange.

Treatment for TB is complex, because (1) the bacteria can spread to many different tissues, and (2) they can develop a resistance to standard antibiotics relatively quickly. As a result, several drugs are used in combination over a period of 6–9 months. The most effective drugs now available include *isoniazid*, which interferes with bacterial replication, and *rifampin*, which blocks bacterial protein synthesis.

Tuberculosis is a major health problem throughout the world, but especially in underdeveloped countries. An estimated 2 billion people are infected at this time. There are 7–9 million cases diagnosed each year and 3 million deaths annually due to tuberculosis infection. The problem is much less severe in developed nations, such as the United States. However, tuberculosis was extremely common in the United States earlier in this century. An estimated 80 percent of the people born around the turn of the century became infected with tuberculosis during their lives. Although many were able to meet the bacterial challenge, it was still the number one cause of death in 1906. These statistics have been drastically altered with the arrival of antibiotics and techniques for early detection of infection. Between 1906 and 1984 the death rate fell from 2 deaths per 1000 population to 1.5 deaths per 100,000 population.

Tuberculosis today is unevenly distributed through the U.S. population, with several groups at relatively high risk for infection. For example, Hispanics, blacks, prison inmates, hospital employees, and individuals with immune disorders, such as AIDS patients, are more likely to be infected than other members of the population. Although at present only 2–5 percent of young American adults have been infected, the incidence and death rate have increased each year since 1984. Recently, strains resistant to many antibiotics have appeared, further complicating efforts to control the disease.

Overloading the Respiratory Defenses   *Page 759*

Large quantities of airborne particles may overload the respiratory defenses and produce a variety of different illnesses. The presence of irritants in the lining of the conducting passageways may provoke the formation of abscesses, and damage to the epithelium in the affected area may allow the irritants to enter the surrounding tissues of the lung. The scar tissue that forms reduces the elasticity of the lung and may restrict airflow along the passageway. Irritants or foreign particles may also enter the lymphatics of the lung, producing inflammation of the regional lymph nodes. Chronic irritation and stimulation of the epithelium and its defenses cause changes in the epithelium that increase the likelihood of lung cancer, discussed further below. (See also Diagnostics: Tissue Structure and Disease on p. 160 of the text, and especially Figure 5-23.) Severe respiratory insufficiency with symptoms of shortness of breath and weight loss develop slowly and may take 20 years or more to appear. **Silicosis** (sil-i-KŌ-sis), produced by the inhalation of silica dust, **asbestosis** (as-bes-TŌ-sis), from the inhalation of asbestos fibers, and **anthracosis** (an-thra-KŌ-sis), the "black lung disease" of coal miners (Figure 23-A), are examples of conditions caused by overloading the respiratory defenses.

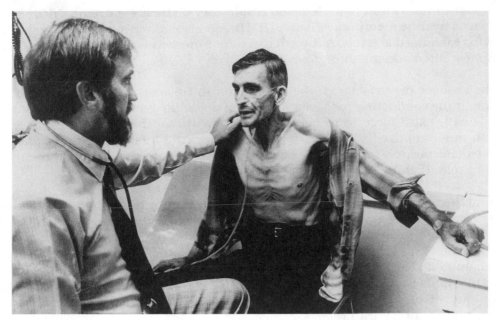

FIGURE 23-A
Miner suffering from black lung disease (anthracosis)

 Thoracentesis *Page 759*

The delicate pleural membranes can become inflamed due to chronic irritation or infection. Inflammation produces symptoms of *pleuritis* or *pleurisy*, conditions introduced in Chapter 5 of the text (see the Clinical Comment: Problems with serous membranes on p. 155). Chronic inflammation often causes a change in the permeability of the pleura, leading to a *pleural effusion* (an abnormal accumulation of fluid within the pleural cavities). When a pleural effusion is detected on an X-ray, samples of pleural fluid are obtained, using a long needle inserted between the ribs. This sampling procedure, introduced in Chapter 8 of this Manual (see The Thoracic Cage and Surgical Procedures) is called *thoracentesis* or *thoracocentesis*. The fluid collected is usually checked for the presence of blood, white blood cells, and bacteria.

Neonatal Respiratory Distress Syndrome (NRDS) *Page 759*

Surfactant cells begin producing surfactants at the end of the sixth fetal month. By the eighth month surfactant production has risen to the level required for normal respiratory function. **Neonatal respiratory distress syndrome** (**NRDS**), also known as **hyaline membrane disease** (**HMD**), develops when surfactant production fails to reach normal levels. Although there are inherited forms of HMD, the condition most often accompanies premature delivery.

In the absence of surfactants, the alveoli tend to collapse during exhalation, and although the conducting passageways remain open, the newborn infant must then inhale with extra force to reopen the alveoli on the next breath. In effect, every breath must approach the power of the first, and the infant rapidly becomes exhausted. Respiratory movements become progressively weaker, eventually the alveoli fail to expand, and gas exchange ceases.

One method of treatment involves assisting the infant by administering air under pressure, so that the alveoli are held open. This procedure, known as **positive end-expiratory pressure** (**PEEP**), can keep the newborn alive until surfactant production increases to normal levels. Surfactant from other sources can also be provided; suitable surfactants can be extracted from cow lungs (*Survanta*), obtained from the fluids that surround full-term infants, or synthesized using gene-splicing techniques (*Exosurf*). These preparations are usually administered in the form of a fine mist of surfactant droplets.

Surfactant abnormalities may also develop in adults, as the result of severe respiratory infections or other sources of pulmonary injury. Alveolar collapse follows, producing a condition known as **adult respiratory distress syndrome** (**ARDS**). PEEP is often used in an attempt to maintain life until the underlying problem can be corrected, but at least 50–60 percent of ARDS cases result in fatalities.

Boyle's Law and Air Overexpansion Syndrome

Page 763

Swimmers descending 4 feet or more beneath the surface experience significant increases in pressure, due to the weight of the overlying water. Air spaces throughout the body decrease in volume. Usually the pressure increase produces mild discomfort in the middle ear, but some people experience acute pain and disorientation. The water pressure first collapses the pharyngotympanic (Eustachian) tubes (see Figures 8-7 and 17-17 on pp. 245 and 543 of the text.) As the volume of air in the middle ear cavities decreases, the tympanic membranes are forced inward. This uncomfortable situation can be remedied by closing the mouth, pinching the nose and exhaling gently, because elevating the pressure in the nasopharynx will force air through the pharyngotympanic tubes and into the middle ear spaces. As the volume of air in each middle ear cavity increases, the tympanic membrane returns to its normal position. When the swimmer returns to the surface, the pressure drops and the air in the middle ear expands. This usually goes unnoticed, for the air simply forces its way along the pharyngotympanic tube and into the nasopharynx.

Scuba divers breathe air under pressure, and that air is delivered at the same pressure as that of their surroundings. A descent to a depth of 33 feet doubles the pressure on a diver, due to the weight of the overlying water. Consider what happens if that diver then takes a full breath of air and heads for the surface. Boyle's law states that as the pressure declines, the volume increases. Thus, at the surface the volume of air in the lungs will have doubled. Such a drastic increase cannot be tolerated, and the usual result is a tear in the wall of the lung. The symptoms and severity of this air overexpansion syndrome depend on where the air ends up. If the air flows into the pleural cavity, the lung will collapse; if it enters the mediastinum, it may compress the pericardium and produce symptoms of cardiac tamponade. Worst of all, the air may rupture blood vessels and enter the circulation. The air bubbles then form emboli that may block blood vessels in the heart or brain, producing a heart attack or stroke. These are all serious conditions, and divers are trained to avoid holding their breath when swimming toward the surface.

Decompression Sickness   *Page 769*

Decompression sickness can develop when an individual experiences a sudden change in pressure. Nitrogen is the gas responsible for this condition. Nitrogen, which accounts for 78.6 percent of the atmospheric gas mixture, has a relatively low solubility in body fluids. Under normal atmospheric pressures there are few nitrogen molecules in the blood, but at higher than normal pressures additional nitrogen molecules diffuse across the alveolar surfaces and into the bloodstream.

As more nitrogen enters the blood it is distributed throughout the body. Over time nitrogen diffuses into peripheral tissues and into body fluids such as the cerebrospinal fluid (CSF), aqueous humor, and synovial fluids. If the pressure decreases, the change must occur slowly enough that the excess nitrogen can diffuse out of the tissues, into the blood, and across the alveolar surfaces. If the pressure falls suddenly, this gradual movement of nitrogen from the periphery to the lungs cannot occur. Instead, the nitrogen leaves solution and forms bubbles of nitrogen gas in the blood, tissues, and body fluids.

A few bubbles in peripheral connective tissues may not be particularly dangerous, at least initially. However, these bubbles can fuse together, forming larger bubbles that distort tissues, causing pain. Bubbles often develop in joint capsules first. These bubbles cause severe pain, and the afflicted individual tends to bend over or curl up. This symptom accounts for the popular name of this condition, "the bends." Bubbles in the systemic or pulmonary circulation can cause infarcts, and those in the cerebrospinal circulation can cause sensory losses, paralysis, or respiratory arrest.

Treatment consists of recompression, exposing the individual to pressures that force the nitrogen back into solution and alleviate the symptoms. Pressures are then reduced gradually over a period of 1 or more days.

Today most bends cases involve scuba divers who have gone too deep or stayed at depth for too great a time. The condition is not restricted to divers, however, and the first reported cases involved construction crews working in pressurized surroundings. Although such accidents are exceedingly rare, the sudden loss of cabin pressure in a commercial airliner can also produce symptoms of decompression sickness.

Bronchitis, Emphysema, and COPD   *Page 771*

Bronchitis (brong-KĪ-tis) is an inflammation of the bronchial lining. The most characteristic symptom is the overproduction of mucus, which leads to frequent coughing. An estimated 20 percent of adult males have chronic bronchitis. This condition is most often related to cigarette smoking, but it can also result from other environmental irritants, such as chemical vapors. Over time the increased mucus production can block smaller airways and reduce respiratory efficiency.

Emphysema (em-fi-SĒ-ma) is a chronic, progressive condition characterized by shortness of breath and an inability to tolerate physical exertion. The underlying problem is destruction of respiratory exchange surfaces. In essence, respiratory bronchioles and alveoli are functionally eliminated. The alveoli gradually expand, and their walls become thickened by fibrous tissue. Capillaries deteriorate, and gas exchange in the region comes to a halt.

Emphysema has been linked to the inhalation of air containing fine particulate matter or toxic vapors, such as those found in cigarette smoke. As the condition progresses, the reduction in exchange surface limits the ability to provide adequate oxygen. The condition is widespread. Some degree of emphysema is a normal consequence of aging, and an estimated 66 percent of adult males and 25 percent of adult females have detectable areas of emphysema in their lungs. Clinical symptoms typically fail to appear unless the damage is extensive.

Chronic bronchitis and emphysema reduce respiratory efficiency, and both conditions are frequently associated with cigarette smoking. The general pattern of symptoms has been called **chronic obstructive pulmonary disease (COPD)** or **chronic airway obstruction (CAO).** Individuals with CAO often expand their chests permanently, in an effort to enlarge their lung capacities and make better use of

100

functional alveoli. This adaptation gives them a distinctive "barrel-chested" appearance.

Two different patterns of symptoms may appear. In one group of patients other aspects of pulmonary structure and function are relatively normal. The respiratory rate in these patients increases dramatically. The lungs are fully inflated at each breath, and expirations are forced. These individuals maintain near-normal arterial pO_2s. Their respiratory muscles are working hard, and they use a lot of energy just breathing. As a result, these patients are usually relatively thin. Because blood oxygenation is near normal, skin color in Caucasian patients will be pink; the combination of rapid, deep breaths and pink skin color has led to the use of the term "pink puffers" for these individuals.

In the second group, emphysema has been complicated by chronic bronchitis. These individuals may also have symptoms of heart failure, including widespread edema. Blood oxygenation is low, and the skin has a bluish coloration. The combination of edema and blue color has led to use of the term "blue bloater" when describing these patients.

The loss of alveoli and bronchioles in emphysema is permanent and irreversible. Further progression can be limited by cessation of smoking; the only effective treatment for acute cases is administration of oxygen.

Blood Gas Analysis   *Page 772*

Blood samples may be analyzed to determine the concentration of dissolved gases. The usual tests include determination of pH, pCO_2, and pO_2 in an arterial sample. This analysis can be very helpful in monitoring patients after a heart attack, COPD, or asthma. A blood sample provides information about the degree of oxygenation in peripheral tissues. For example, if the pCO_2 is very high and the pO_2 very low, tissues are not receiving adequate oxygen. This condition may be corrected by providing a gas mixture that has a high pO_2 (or even pure oxygen, with a pO_2 of 760 mm Hg). In addition, these measurements determine the efficiency of gas exchange at the lungs. If the arterial pO_2 remains low despite oxygen administration, or the pCO_2 is very high, pulmonary exchange problems must exist, such as pulmonary edema, emphysema, or pneumonia.

Carbon Monoxide Poisoning   *Page 773*

Mystery writers often describe murder or suicide victims who died in their cars inside a locked garage. In real life, entire families are killed each winter by leaky furnaces or space heaters. The cause of death is *carbon monoxide poisoning*. The exhaust of automobiles, other petroleum-burning engines, oil lamps, and fuel-fired space heaters contain **carbon monoxide** (CO). Carbon monoxide competes with oxygen molecules for the binding sites on heme units. Unfortunately, the carbon monoxide usually wins, for it has a much stronger affinity for hemoglobin at very low partial pressures. The bond is extremely durable, and the attachment of a carbon monoxide molecule essentially inactivates that heme unit for respiratory purposes. If carbon monoxide molecules make up 0.1 percent of the components of inspired air, enough hemoglobin will be affected that survival will become impossible without medical assistance. Treatment includes breathing pure oxygen, for under high partial pressures the oxygen molecules will "bump" the carbon monoxide from the hemoglobin.

 ## Chemoreceptor Accommodation and Opposition

Page 779

Carbon dioxide receptors show accommodation to sustained pCO_2 levels above or below normal. Although they register the initial change quite strongly, over a period of days they adapt to the new level as "normal." As a result, after several days of elevated pCO_2 levels the effects on the respiratory centers begin to decline. Fortunately the response to a low arterial pO_2 remains intact, and the response to arterial oxygen concentrations becomes increasingly important.

Some patients with severe chronic lung disease are unable to maintain normal pO_2 and pCO_2 levels in the blood. The arterial pCO_2 rises to 50–55 mm Hg, and the pO_2 falls to 45–50 mm Hg. At these levels the carbon dioxide receptors accommodate, and most of the respiratory drive comes from the arterial oxygen receptors. If these patients are given too much oxygen, they may simply stop breathing! Vigorous stimulation to encourage breathing or a mechanical respirator may be required.

Shallow Water Blackout

Page 779

Many swimmers attempt to outwit the chemoreceptor reflexes and extend their time under water. The usual method involves taking a number of extra-deep breaths before submerging. These individuals are intentionally hyperventilating, usually with the stated goal of "taking up extra oxygen."

From the hemoglobin saturation curve it should be obvious that this explanation is totally incorrect. At a normal alveolar pO_2 the hemoglobin will be 97.5 percent saturated, and no matter how many breaths the swimmer takes the alveolar pO_2 will not rise significantly. But the pCO_2 *will* be affected because the increased ventilation rate lowers the carbon dioxide concentrations of the alveoli and blood. This produces the desired effect by temporarily shutting off the chemoreceptors monitoring the pCO_2. As long as CO_2 levels remain depressed, the swimmer does not feel the "need" to breathe, despite the fact that pO_2 continues to fall.

Under normal circumstances, breath holding causes a decline in pO_2 and a rise in pCO_2. As indicated in Figure 23-29, by the time the pO_2 has fallen to 60 percent of normal levels, carbon dioxide levels will have risen enough to make breathing an unavoidable necessity. But following hyperventilation, oxygen levels can fall so low that the swimmer becomes unconscious (usually at a pO_2 of 15–20 mm Hg) before the pCO_2 rises enough to stimulate breathing. Because this "shallow water blackout" usually has fatal consequences, *hyperventilation should be avoided by swimmers and divers.*

Sudden Infant Death Syndrome (SIDS)   *Page 780*

Sudden infant death syndrome (SIDS), also known as "crib death," kills an estimated 10,000 infants each year in the United States alone. Most crib deaths involve infants 2–4 months old, usually between midnight and 9:00 A.M., in the late fall or winter months. Eyewitness accounts indicate that the sleeping infant suddenly stops breathing, turns blue, and relaxes. There appear to be genetic factors involved, but controversy remains as to the relative importance of other factors, such as laryngeal spasms, cardiac arrhythmias, upper respiratory tract infections, viral infections, and/or CNS malfunctions.

CHAPTER 24

The Digestive System

Dental Problems *Page 792*

Food particles remaining in the mouth after a meal provide a buffet for oral bacteria, and most common dental problems result from the action of these bacteria. Bacteria adhering to the surfaces of the teeth produce a sticky matrix that traps food particles and creates deposits of **plaque**. The mass of the plaque deposit protects the bacteria from salivary secretions, and as they digest the nutrients these oral pests generate acids that gradually erode the structure of the tooth. The results are **dental caries**, otherwise known as *cavities*. Brushing the exposed surfaces of the teeth after meals helps to prevent the settling of bacteria and the entrapment of food particles, but bacteria between the teeth and within the gingival sulcus may elude the brush. Dentists therefore recommend the use of dental floss to clean these spaces daily.

If the bacteria remain unchallenged within the gingival sulcus, the acids generated begin eroding the connections between the neck of the tooth and the gingiva. The gums appear to recede from the teeth, and **periodontal disease** develops. As it progresses the bacteria attack the cementum, progressively destroying the periodontal ligament and eroding the alveolar bone. This deterioration loosens the tooth, and periodontal disease is the most common cause for the loss of teeth. If the bacteria reach the pulp and infect it, **pulpitis** (pul-PĪ-tis) results. Treatment usually involves the complete destruction of the pulp cavity and the afferent nerve, followed by packing the tooth with solid filling materials. This procedure is called a **root canal** therapy.

If teeth are broken or they must be removed due to disease, the usual treatment involves replacing them with "false teeth" attached to a plate or frame inserted into the mouth. Over the past 10 years an alternative has been developed, using **dental implants**. A ridged titanium cylinder is inserted into the alveolus, and osteoblasts lock the ridges into the surrounding bone. After 4–6 months, an artificial tooth is screwed into the cylinder. It is estimated that roughly 300,000 dental implants will have been performed in the United States by 1992. The use of dental implants will almost certainly increase substantially over the next decade, as the "baby boomers" age. Roughly 42 percent of individuals over age 65 have lost all of their teeth; the rest have lost an average of 10 teeth.

Achalasia and Esophagitis

Page 795

In the condition known as **achalasia** (ak-a-LĀ-zē-a), a bolus descends relatively slowly and its arrival does not cause the opening of the lower esophageal sphincter. Materials then accumulate at the base of the esophagus like cars at a stop light. Secondary peristaltic waves may occur repeatedly, adding to the individual's discomfort. The most successful treatment involves cutting the circular muscle layer at the base of the esophagus or expanding a balloon in the lower esophagus until the muscle layer tears.

A weakened or permanently relaxed sphincter can cause inflammation of the esophagus, or **esophagitis** (ē-sof-a-JĪ-tis), as powerful gastric acids enter the lower esophagus. The esophageal epithelium has few defenses from acid and enzyme attack, and inflammation, epithelial erosion, and intense discomfort are the result. Occasional incidents of reflux, or backflow, from the stomach are responsible for the symptoms of "heartburn." This relatively common problem supports a multimillion dollar industry devoted to producing and promoting antacids.

Gastritis and Peptic Ulcers

Page 798

Inflammation of the gastric mucosa causes **gastritis** (gas-TRĪ-tis). This condition may develop after swallowing drugs, including alcohol and aspirin. Gastritis may also appear after severe emotional or physical stress, bacterial infection of the gastric wall, or the ingestion of strongly acid or alkaline chemicals.

A **peptic ulcer** develops when the digestive acids and enzymes manage to erode their way through the defenses of the stomach lining or proximal portions of the small intestine. The locations may be indicated by using the terms **gastric ulcer** (stomach) or **duodenal ulcer** (duodenum). Peptic ulcers result from the excessive production of acid or the inadequate production of the alkaline mucus that poses an epithelial defense.

Recent evidence suggests that infections involving the bacterium *Helicobacter pylori* are an important factor in ulcer formation. These bacteria are able to survive long enough to penetrate the mucus coating the epithelium. Once within the protective layer of mucus, they are safe from the action of gastric acids and enzymes. Over time, the infection damages the epithelial lining, with two major results: (1) erosion of the lamina propria by gastric juices and (2) entry and spread of the bacteria through the gastric wall and into the bloodstream. Recommended treatment involves at least 2–3 weeks of therapy using a combination of antibiotics.

The importance of bacterial infection in peptic ulcers remains controversial. One group of researchers is convinced that all ulcers are the result of bacterial infection and that triple-drug treatment could cure 90 percent of all cases. Others accept the importance of bacterial infection in ulcer production, but do not think that bacteria are always involved. However, everyone now realizes that these bacteria can cause problems outside of the digestive tract as well as inside. At least seven patients are known to have died after receiving tranfusions of blood from donors with peptic ulcers. The deaths resulted from massive systemic infections traced to the presence of bacteria in the donor's blood.

Regardless of the primary cause, once gastric juices have destroyed the epithelial layers, the virtually defenseless lamina propria will be exposed to digestive attack. Sharp abdominal pain results, and bleeding can develop. The administration of antacids can often control peptic ulcers by neutralizing the acids and allowing time for the mucosa to regenerate. The drug **cimetidine** (sī-MET-i-dēn), or *Tagamet*, inhibits the secretion of acid by the parietal cells. Dietary restrictions limit the intake of acidic beverages and eliminate foods that promote acid production (caffeine) or damage unprotected mucosal cells (alcohol). In severe cases the damage may provoke significant bleeding, and the acids may even erode their way through the wall of the digestive tract and enter the peritoneal cavity. This condition, called a **perforated ulcer**, requires immediate surgical correction.

Stomach Cancer *Page 799*

Stomach cancer is one of the most common lethal cancers, responsible for roughly 15,000 deaths in the United States each year. Because the symptoms may resemble those of gastric ulcers, the condition may not be reported in its early stages. Diagnosis usually involves X-rays of the stomach at various degrees of distension. (See Figure 1-14 on p. 20 of the text). The mucosa can also be visually inspected using a flexible instrument called a **gastroscope**. Attachments permit the collection of tissue samples for histological analysis.

Treatment of gastric cancer involves the surgical removal of part or all of the stomach. Even a total **gastrectomy** (gas-TREK-to-mē) can be tolerated because the stomach provides no essential digestive services other than the secretion of intrinsic factor.

Drastic Weight-Loss Techniques *Page 802*

At any given moment, an estimated 20 percent of the U.S. population is dieting to promote weight loss. (The relationships between fad diets, nutrition, and common sense are discussed in Chapter 25 of the text—see especially the Health News box on pp. 856–57.) In addition to the appearance of "fat farms" and exercise clubs across the country, there has been an increase in the use of surgery to promote weight loss. Many of these techniques involve surgically remodeling the gastrointestinal tract. **Gastric stapling** attempts to correct an overeating problem by reducing the size of the stomach. A large portion of the gastric lumen is stapled closed, leaving only a small pouch in communication with the esophagus and duodenum. After this surgery the individual will be able to eat only a small amount before the stretch receptors in the gastric wall become stimulated and the individual feels "full."

Gastric stapling is a major surgical procedure, and there are many potential complications. In addition, the smooth muscle in the wall of the functional portion of the stomach gradually becomes increasingly tolerant of distension, and the operation may have to be repeated.

A more drastic approach involves the surgical removal or bypass of a large portion of the jejunum. This procedure reduces the effective absorptive area, producing a marked weight loss. After the operation the individual must follow a very restricted diet and take a number of dietary supplements to ensure that all the essential nutrients and vitamins can be absorbed before the chyme enters the large intestine. Chronic diarrhea and serious liver disease are potential complications of this procedure.

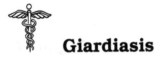

Giardiasis   *Page 803*

Giardiasis is an infection caused by the protozoan *Giardia lamblia*. This pathogen colonizes the duodenum and jejunum and interferes with the normal absorption of lipids and carbohydrates. Many people do not develop acute symptoms, and these individuals act as carriers who can spread the disease. When acute symptoms develop, they usually appear within 3 weeks of initial exposure. Violent diarrhea, abdominal pains, cramps, nausea, and vomiting are the primary complaints. These symptoms persist for 5–7 days, although some patients are subject to relapses. Treatment usually consists of oral administration of drugs such as *quinacrine* or *metronidazole* that can kill the protozoan.

Transmission of giardiasis requires the contamination of food or water with feces containing *cysts*, resting stages of the protozoan that are produced during passage through the large intestine. Rates of infection are highest in Third World countries with poor sanitation, among individuals with impaired immune systems (as in AIDS), and among toddlers and young children. The cysts can survive in the environment for months, and they are not killed during the chlorine treatment used to kill bacteria in drinking water. Travelers are advised to boil water or food before consumption, as this effectively kills the cysts.

Vomiting and Intestinal Evacuation *Page 803*

The responses of the digestive tract to chemical or mechanical irritation are usually quite predictable. The production of fluids increases, and the contents are eliminated as quickly as possible. The **vomiting reflex** occurs in response to irritation of the digestive lining. The fauces and pharynx are particularly responsive to unpleasant stimuli, but vomiting may also be caused by irritation of the esophagus, stomach, or proximal portions of the small intestine. During the **preparatory phase**, or *pre-emetic phase*, the individual salivates heavily and has a sensation of nausea. Over this period the pylorus relaxes, and the contents of the duodenum and proximal jejunum are discharged into the stomach by strong peristaltic waves that travel "in reverse." Vomiting, or **emesis** (EM-e-sis), then occurs as the stomach regurgitates its contents through the esophagus and pharynx.

Most of the force of vomiting comes from a forceful expiratory movement that elevates intra-abdominal pressures. Coordination for this reflex occurs under the direction of a **vomiting center** in the medulla. Emotional or sensory stimuli may also lead to a vomiting response by influencing this center. Drugs administered to promote vomiting are called **emetics** (i-MET-iks). *Emetine*, or *ipecac*, is a strongly emetic drug that induces a combination of nausea, vomiting, and diarrhea.

Gastroenteritis *Page 803*

An irritation of the small intestine may lead to a series of powerful peristaltic contractions that eject the contents of the small intestine into the large intestine. An extremely powerful irritating stimulus will produce a "clean sweep" of the absorptive areas of the digestive tract. Vomiting clears the stomach, duodenum, and proximal jejunum, and peristaltic contractions evacuate the distal jejunum and ileum. Bacterial toxins, viral infections, and various poisons will sometimes produce these extensive gastrointestinal responses. Conditions affecting primarily the small intestine are usually referred to as **enteritis** (en-ter-Ī-tis) of one kind or another. If both vomiting and diarrhea are present, the term **gastroenteritis** (gas-trō-en-ter-Ī-tis) may be used instead.

Diverticulitis and Colitis *Page 806*

In **diverticulosis** (dī-ver-tik-ū-LŌ-sis) pockets (*diverticula*) form in the mucosa, usually in the sigmoid colon. These get forced outward, probably by the pressures

108

generated during defecation. If the pockets push through weak points in the muscularis externa, they form semi-isolated chambers that are subject to recurrent infection and inflammation. The infections cause pain and the occasional bleeding characteristic of **diverticulitis** (dī-ver-tik-ū-LĪ-tis). In severe cases the diverticula may perforate, setting the bacteria loose in the peritoneal cavity.

The general term **colitis** (ko-LĪ-tis) may be used to indicate a condition characterized by inflammation of the colon. The **irritable bowel syndrome** is characterized by diarrhea, constipation, or an alternation between the two. When constipation is the primary problem, this condition may be called a *spastic colon*, or *spastic colitis*. The irritable bowel syndrome may have a psychological basis. **Inflammatory bowel disease**, or *ulcerative colitis*, involves chronic inflammation of the digestive tract, most often affecting the colon. The mucosa becomes inflamed and ulcerated, extensive areas of scar tissue develop, and colonic function deteriorates. Acute diarrhea, cramps, and bleeding are common symptoms. Fever and weight loss are also frequent complaints. Treatment of inflammatory bowel disease often involves anti-inflammatory drugs and corticosteroids that reduce inflammation. In severe cases, oral or intravenous fluid replacement is required. In cases that do not respond to other therapies, immunosuppressive drugs, such as *cyclosporine* (see p. 745 of the text) may be used to good effect.

Treatment of severe inflammatory bowel disease may also involve a **colectomy** (kō-LEK-to-mē), the removal of all or a portion of the colon. If a large part or even all of the colon must be removed, normal connection with the anus cannot be maintained. Instead, the end of the intact digestive tube is sutured to the abdominal wall, and wastes then accumulate in a plastic pouch or sac attached to the opening. If the attachment involves the colon, the procedure is a **colostomy** (ko-LOS-to-mē); if the ileum is involved it is an **ileostomy** (il-ē-OS-to-mē).

Diarrhea and Constipation *Page 810*

Diarrhea (dī-a-RĒ-a) results when the ileal and colonic mucosa becomes unable to maintain normal levels of absorption or if the rate of fluid entry exceeds its maximum capacity. The result is the copious production and defecation of watery feces.

Bacterial, viral, or protozoan infection of the colon or small intestine may cause acute bouts of diarrhea lasting several days. One example, the "traveler's trots," sometimes afflicts those on vacation in foreign countries. The condition, usually caused by a bacterial or viral infection, develops because the irritated or damaged mucosal cells are unable to maintain normal absorption levels. The irritation stimulates the production of mucus, and the damaged cells and mucus secretions add to the volume of feces produced.

Despite the inconvenience, most of these conditions are temporary, and mild diarrhea is probably a reasonably effective method of rapidly removing an intestinal irritant. Drugs, such as *Lomotil*, that prevent peristaltic contractions in the colon relieve the diarrhea but leave the irritant intact, and the symptoms may return with a vengeance when the drug effects fade.

Severe diarrhea can be life-threatening, due to cumulative fluid and ion losses. In **cholera** (KOL-e-ra), bacteria bound to the intestinal lining release toxins that stimulate a massive fluid secretion across the intestinal epithelium. Fluid loss during the worst stage of the disease can approach 1 liter per hour. This dramatic loss causes a rapid drop in blood volume, leading to acute hypovolemic shock and damage to the kidneys and other organs. Without treatment the victim may die of acute dehydration in a matter of hours. Cholera epidemics are most common in areas with poor sanitation and where drinking water is contaminated by fecal wastes. After an incubation period of 1–2 days, the symptoms of nausea, vomiting, and diarrhea persist for 2–7 days. Treatment consists of oral or intravenous fluid replacement while the disease runs its course. There is a vaccine available, but its low success rate (40–60 percent) and short duration (4–6 month protection) make it relatively ineffective in preventing or controlling cholera outbreaks. In 1991, a cholera epidemic began in Peru and has since spread throughout South America. By early 1992 more than 500,000 cases had been reported, with a death rate of 0.5 percent. This outbreak has had a remarkably low mortality rate; death rates in other outbreaks in this century have been as high as 60 percent.

Constipation occurs when fecal materials are moving through the colon so slowly that excessive water resorption occurs. The feces then become dry, hard, abrasive, and difficult to move. Inadequate dietary fiber and fluids, coupled with a lack of exercise, are the usual causes. Indigestible fiber adds bulk to the feces, retaining moisture and stimulating stretch receptors that promote peristalsis. Active movement during exercise assists in the movement of fecal materials through the colon.

Constipation can usually be treated by oral administration of stool softeners, such as *Colace*, **laxatives**, or **cathartics** (ka-THAR-tiks) that promote defecation. These compounds promote water movement into the feces, increase fecal mass, or irritate the lining of the colon to stimulate peristalsis.

 **Mumps** *Page 810*

The **mumps** virus preferentially targets the salivary glands, most often the parotids, although other organs may also become infected (Figure 24-A). Infection most often occurs at 5–9 years of age. The first exposure stimulates antibody production and usually confers permanent immunity; active immunity can be conferred by

110

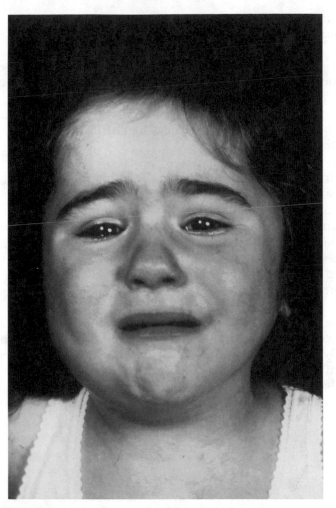

FIGURE 24-A
Swollen salivary glands typical of mumps

immunization. In postadolescent males the mumps virus may also infect the testes and cause sterility. Infection of the pancreas by the mumps virus may produce temporary or permanent diabetes; other organ systems, including the CNS, may be affected in severe cases.

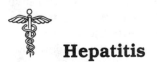 **Hepatitis** *Page 814*

Viruses specifically targeting the liver are responsible for several forms of **hepatitis** (hep-a-TĪ-tis). **Hepatitis A**, or "infectious hepatitis," usually results from ingestion of food, water, milk, or shellfish contaminated by fecal wastes. It has a relatively short incubation period of 2–6 weeks.

Hepatitis B, or "serum hepatitis," is transmitted by intimate contact. For example, infection may occur through blood products, via a break in the skin or mucosa, or by sexual contact. The incubation period is much longer, ranging from 1 to 6 months. There appear to be several other hepatitis viruses, including one originally designated "non-A, non-B" and now called **hepatitis C**. These forms of hepatitis are as yet poorly understood.

The hepatitis viruses disrupt liver function by attacking liver cells. The individual develops a chronic fever, and the liver may become inflamed and tender. Several hematological parameters change markedly. For example, enzymes usually confined to the cytoplasm of functional liver cells appear in the circulating blood. Normal metabolic regulatory activities become less effective, and blood glucose declines. Plasma protein synthesis slows, and the clotting time becomes unusually long. The injured hepatocytes stop removing bilirubin from the circulating blood, and symptoms of jaundice appear.

Almost everyone who contracts viral hepatitis eventually recovers, although full recovery may take several months. About 10 percent of hepatitis B patients develop potentially dangerous complications. Passive immunization with pooled immunoglobulins is available for the hepatitis A and B viruses. (Active immunization for hepatitis B is also available.) Hepatitis with comparable symptoms may also result from exposure to toxic chemicals and drugs. These may affect only the liver, or the liver may be one of many organs under attack.

 **Cirrhosis** *Page 815*

The underlying problem in **cirrhosis** (sir-Ō-sis) appears to be the widespread destruction of hepatocytes by exposure to drugs (especially alcohol), viral infection, ischemia, or blockage of the hepatic ducts. Two processes are involved in producing the symptoms. Initially the damage to hepatocytes leads to the formation of extensive areas of scar tissue that branch throughout the liver. The surviving hepatocytes then undergo repeated cell divisions, but the fibrous tissue prevents the new hepatocytes from achieving a normal lobular arrangement. So the liver gradually converts from an organized assemblage of lobules to a fibrous aggregation of poorly functioning cell clusters. Jaundice, ascites, and other symptoms may appear as the condition progresses.

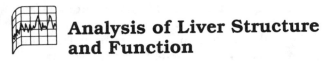

Analysis of Liver Structure and Function

Page 815

A variety of clinical tests are used to check the functional and physical state of the liver. **Liver function tests** assess specific functional capabilities. A **serum bilirubin assay** indicates how efficiently the liver has been able to extract and excrete this compound. Serum and plasma protein assays can detect changes in the liver's rate of plasma protein synthesis, and serum enzyme tests can reveal liver damage by detecting intracellular enzymes in the circulating blood.

Liver scans involve the injection of radioisotope-labeled compounds into the circulation. Compounds are chosen that will be selectively absorbed by either Kupffer cells, liver cells, or abnormal liver tissues. CT scans are often used to provide information about cysts, abscesses, tumors, or hemorrhages in the liver.

In **laparoscopy** (lap-a-ROS-ko-pē) a flexible fiber optic instrument introduced through the abdominal wall permits direct visualization of the viscera and can be used to take a tissue sample for histological analysis. A liver biopsy can also be taken through the abdominal wall using a long needle.

Problems with Bile Storage and Secretion

Page 815

If bile becomes *too* concentrated, crystals of insoluble minerals and salts begin to appear. These deposits are called **gallstones**. Merely having them, a condition termed **cholelithiasis** (kō-lē-li-THĪ-a-sis; *chole*, bile), does not represent a problem as long as the stones remain small. An estimated 16–20 million people in the United States have gallstones that are unnoticed, and small stones are often flushed down the bile duct and excreted.

If gallstones enter and jam in the cystic or bile duct, the painful symptoms of **cholecystitis** (kō-lē-sis-TĪ-tis) appear. About a million people develop acute symptoms of cholecystitis each year. The gallbladder becomes swollen and inflamed, infections may develop, and if the blockage does not work its way down the duct to the duodenum it must be removed or destroyed. Small gallstones can be chemically dissolved. One chemical now under clinical review is *methyl tert-butyl ether* (*MTBE*). When introduced into the gallbladder it dissolves gallstones in a matter of hours; testing to date has not shown many undesirable side effects. (You may have heard of this compound in another context, as it is a gasoline additive used to prevent engine knocking.)

Surgery is usually required to remove large gallstones. (The gallbladder is also removed to prevent recurrence.) Some surgeons are now using a laparoscope (see Endometriosis in Chapter 29 of this Manual), inserted through a very small incision, to perform this surgery. A newly developed therapy involves immersing the individual in water and shattering the stones with focused sound waves. The particles produced are then small enough to pass through the bile duct without difficulty.

The bile duct or hepatic duct may also be blocked by external compression, parasitic infection, or tumor formation. In all of these conditions, bile excretion comes to an abrupt halt. As a result, pressure builds in the tributaries of the bile duct, and eventually the hepatocytes become unable to secrete any more bile. At this point the bilirubin concentration in the blood begins to rise. This leads to the symptoms of *obstructive jaundice*, a condition introduced in Chapter 19 of this Manual (see Bilirubin Tests and Jaundice).

Recurrent inflammation and engorgement of the gallbladder may ultimately lead to degenerative changes. Leakage or actual rupture of the gallbladder may then release bile into the abdominal cavity, and the gallbladder must be surgically removed. This loss does not seriously impair digestion, for bile production continues at normal levels.

Pancreatitis   *Page 817*

Pancreatitis (pan-krē-a-TĪ-tis) is an inflammation of the pancreas. Blockage of the excretory ducts, bacterial or viral infections, circulatory ischemia, and drug reactions, especially those involving alcohol, are among the factors that may produce this condition. These stimuli provoke a crisis by injuring exocrine cells in at least a portion of the organ. Lysosomes within the damaged cells then activate the proenzymes, and **autodigestion** begins. The proteolytic enzymes digest the surrounding, undamaged cells, activating their enzymes and starting a chain reaction.

In most cases only a portion of the pancreas will be affected, and the condition subsides in a few days. In 10–15 percent of pancreatitis cases, the process does not subside, and the enzymes may ultimately destroy the pancreas. They are also likely to enter the systemic circulation, via the pancreatic blood vessels, and digest their way through the pancreatic capsule and enter the abdominopelvic cavity. Roughly two-thirds of these patients survive, but they often suffer recurrent painful episodes of pancreatitis.

Malabsorption Syndromes   *Page 825*

Difficulties in the absorption of all classes of compounds will result from damage to the accessory glands or the intestinal mucosa. If the accessory organs are func-

114

tioning normally, but their secretions cannot reach the duodenum, the condition is called **biliary obstruction** (bile duct blockage) or **pancreatic obstruction** (pancreatic duct blockage). Alternatively, the ducts may remain open but the glandular cells are damaged and unable to continue normal secretory activities. Two examples, *pancreatitis* and *cirrhosis* of the liver, were discussed earlier in the chapter.

Even with the normal enzymes present in the lumen, absorption will not occur if the mucosa cannot function properly. A genetic inability to manufacture certain key enzymes will result in discrete patterns of malabsorption—*lactose intolerance* is a good example. Mucosal damage due to ischemia, radiation exposure, toxic compounds, or infection will affect absorption in general and will deplete nutrient and fluid reserves as a result.

CHAPTER 25

Metabolism and Energetics

Phenylketonuria   *Page 834*

Phenylketonuria (fen-il-kē-tō-NU-rē-a), or **PKU**, is one of about 130 disorders that have been traced to the lack of a specific enzyme. Individuals suffering from PKU are deficient in a key enzyme responsible for the conversion of the amino acid phenylalanine to tyrosine. This reaction is a necessary step in the synthesis of tyrosine, an important component of many proteins and the structural basis for a pigment (melanin), two hormones (epinephrine and norepinephrine), and two neurotransmitters (dopamine and norepinephrine). This conversion must also occur before the carbon chain of a phenylalanine molecule can be recycled or broken down in the TCA cycle.

If PKU is undetected and untreated, plasma concentrations of phenylalanine gradually escalate from normal (about 3 mg/dℓ) to levels above 20 mg/dℓ. High plasma concentrations of phenylalanine affect overall metabolism, and a number of unusual byproducts are excreted in the urine. The synthesis and degradation of proteins and other amino acid derivatives are affected. Developing neural tissue is most strongly influenced by these metabolic alterations, and severe brain damage results.

Fortunately, this condition is detectable shortly after birth because there are elevated levels of phenylalanine in the blood and *phenylketone*, a metabolic byproduct, in the urine of the infant. Treatment consists of controlling the amount of phenylalanine in the diet while monitoring plasma concentrations. Individuals with PKU must carefully monitor the ingredients used in the preparation of their meals. For example, one popular artificial sweetener, *Nutrasweet®*, consists of phenylalanine and aspartic acid. Consumption of food or beverages containing this sweetener can therefore cause severe problems for PKU sufferers. Because tyrosine cannot be synthesized from dietary phenylalanine, the diet of these patients must also contain adequate amounts of tyrosine.

In its most severe form PKU affects approximately 1 infant in 20,000. Individuals who carry only a single gene for PKU will produce the affected enzyme but in lesser amounts. They are asymptomatic but have slightly elevated phenylalanine levels in their blood. Statistical analysis of the incidence of fully developed PKU indicates that as many as 1 person in 70 may carry a gene for this condition.

Protein Deficiency Diseases *Page 835*

Protein deficiency diseases develop when an individual does not consume adequate amounts of all essential amino acids. *All* amino acids must be available if protein synthesis is to occur. Every tRNA must appear at the proper location bearing its individual amino acid; as soon as the amino acid called for by a particular codon is missing, the entire process comes to a halt. *Regardless of the energy content of the diet, if it is deficient in essential amino acids the individual will be malnourished to some degree.*

In a protein deficiency disease protein synthesis decreases throughout the body. As protein synthesis in the liver fails to keep pace with the breakdown of plasma proteins, plasma osmolarity falls. This reduced osmolarity results in a fluid shift, as more water moves out of the capillaries and into interstitial spaces, the peritoneal cavity, or both. The longer the individual remains in this state, the more severe the ascites and edema that results. (Ascites and liver function are discussed in the Clinical Comment: Portal Hypertension on p. 815 of the text.)

This clinical picture is relatively common outside of the United States, where dietary protein is often scarce or prohibitively expensive. Growing infants suffer from **marasmus** (ma-RAZ-mus) when deprived of adequate proteins and calories. **Kwashiorkor** (kwash-ē-ŌR-kor) occurs in children whose protein intake is inadequate, even if the caloric intake is acceptable (Figure 25-A). In each case, additional complications include damage to the developing brain. It is estimated that over 100 million children worldwide are suffering from protein deficiency diseases.

FIGURE 25-A
Kwashiorkor

Gout   *Page 837*

Normal plasma uric acid concentrations average 2.7–7.4 mg/dℓ, depending on sex and age. When plasma concentrations exceed 7.4 mg/dℓ, **hyperuricemia** (hī-per-ū-ri-SĒ-mē-a) exists. This condition may affect 18 percent of the U.S. population.

At these concentrations body fluids are supersaturated with uric acid. Although symptoms may not appear at once, uric acid crystals begin to precipitate in body fluids. The condition that then develops is called **gout**. Its severity is determined by the amount and location of the precipitate.

117

Initially the joints of the extremities, especially the metatarsal/phalangeal joint of the big toe, are likely to be affected. This exquisitely painful condition, called **gouty arthritis**, may persist for several days and then disappear for a period of days to years. Recurrences often involve other joints as well as producing generalized fevers. Precipitates may also form within other cartilages, in synovial fluids, tendons, or other connective tissues, or in the kidneys. At serum concentrations of over 12–13 mg/dℓ, half of the patients will develop kidney stones, and kidney function may be affected to the point of kidney failure.

The incidence of gout is much lower than that of hyperuricemia, with estimates ranging from 0.13 to 0.37 percent of the population. Only about 5 percent of the sufferers are women, and most affected males are over 50. Foods high in purines, such as meats or fats, may aggravate or initiate the onset of gout. These foods usually cost more than carbohydrates, and "rich foods" have often been associated with this condition.

 **Obesity** *Page 839*

A normal individual retains around 2 months' supply of energy in the triglycerides of adipose tissue. The evolutionary advantages are obvious, in providing a buffer against daily, monthly, and even seasonal changes in the available food supply. But as long as an individual continues to take in more nutrients than he or she can catabolize or excrete, the excess will continue to be stored as fat. Most of the increase in triglyceride mass represents an increase in the size of individual adipocytes. An increase in the total number of adipocytes does not ordinarily occur, except in children before puberty and in extremely obese adults.

How fat does a person have to be to qualify as obese? A useful definition is 20 percent over ideal weight, as this is the point at which serious health risks appear. Using that criterion, 20–30 percent of men and 30–40 percent of women in the United States can be considered obese!

Simply stated, obese individuals are taking in more food energy than they are using. Unfortunately there is very little agreement as to the underlying cause for this situation. There appear to be two major categories of obesity: *regulatory obesity* and *metabolic obesity*.

Regulatory obesity results from a failure to regulate food intake so that appetite, diet, and activity are in balance. Most instances of obesity fall within this category. Usually there are no obvious organic causes.[1] The chronic overeating may result

---

[1]In rare cases the problem may arise because of some disorder, such as a tumor, affecting the hypothalamic centers dealing with appetite and satiation.

118

from psychological or sociological factors, such as stress, neurosis, long-term habits, family or ethnic traditions, or from inactivity. Genetic factors may also be involved, but because the psychological and social environment plays such an important role in human behavior the exact connections have been difficult to assess. In short, individuals suffering from regulatory obesity are overeating for some reason and thereby extending the duration and magnitude of the absorptive state.

In **metabolic obesity** the condition is secondary to some underlying organic malfunction that affects cell and tissue metabolism. These cases are relatively rare and typically involve chronic hypersecretion or hyposecretion of metabolically active hormones, such as glucocorticoids.

In a clinical setting, categorizing an obesity problem is less important than determining the degree of obesity and the number and severity of the related complications. The affected individuals are at a much greater risk of developing diabetes, hypertension, and coronary artery disease, as well as gallstones, thromboemboli, hernias, arthritis, varicose veins, and some forms of cancer. A variety of treatments may be considered, ranging from behavior modification counseling or psychotherapy to gastric stapling or the bypass of a portion of the jejunum.

Iron Deficiencies and Excesses *Page 847*

The body of a normal man contains around 3.5 g of iron in the ionic form Fe^{2+}. Of that amount, 2.5 g are bound to the hemoglobin of circulating red blood cells, and the rest is stored in the liver and bone marrow. In women, the total body iron content averages 2.4 g, with roughly 1.9 g incorporated into red blood cells. Thus a woman's iron reserves consist of only 0.5 g, half that of a typical man.

Any condition that produces a blood loss reduces the body iron content. As the lost red blood cells are replaced, iron reserves must be mobilized for use in the synthesis of new hemoglobin molecules. If those reserves are exhausted, or dietary sources are inadequate, symptoms of **iron deficiency** appear. In this condition red blood cells are unable to synthesize functional hemoglobin, and they are unusually small when they enter the circulation. The hematocrit declines, and the hemoglobin content and oxygen-carrying capacity of the blood are substantially reduced. The combination of low hematocrit, low hemoglobin content, and reduced oxygen-carrying capacity is called **anemia**. There are many different forms of anemia. **Iron deficiency anemia** results from inadequate iron reserves. Symptoms include weakness and a tendency to fatigue easily.

Because their reserves are relatively small, women are dependent on a reliable dietary supply of iron. When the demand for iron increases out of proportion with dietary supplies, iron deficiency develops. An estimated 20 percent of menstruating women in the United States show signs of iron deficiency. Pregnancy also stresses iron reserves, for the woman must provide the iron needed to produce both maternal and fetal erythrocytes.

Good dietary sources of iron include liver, red meats, kidney beans, egg yolks, spinach, and carrots. Iron supplements can help prevent iron deficiency, but too much iron can be as dangerous as too little. Iron absorption across the digestive tract normally keeps pace with physiological demands. When the diet contains abnormally high concentrations of iron, or hereditary factors increase the rate of absorption, the excess iron gets stored in peripheral tissues. Eventually cells begin to malfunction as massive iron deposits accumulate in the cytoplasm. For example, iron deposits in pancreatic cells can lead to diabetes mellitus; deposits in cardiac muscle fibers lead to abnormal heart contractions. Liver cells become nonfunctional, and liver cancers may develop.

Comparable symptoms of "iron loading" may appear following repeated transfusions of whole blood, because each unit of whole blood contains roughly 250 mg of iron. For example, the various forms of **thalassemia** result from a genetic inability to produce adequate amounts of one of the four globin chains in hemoglobin. Erythrocyte production and survival are reduced, and so is the oxygen-carrying capacity of the blood. Individuals with severe untreated thalassemia usually die in their twenties, but *not because of the anemia*. These patients are treated for severe anemia with frequent blood transfusions, and the excessive iron loading eventually leads to fatal heart problems.

Hypervitaminosis   *Page 848*

"If a little is good, a lot must be better" is a common but dangerously incorrect attitude about vitamins. Americans spend billions of dollars each year purchasing vitamins, vitamin supplements, and "natural" vitamins that are actually identical to the "synthetic" forms. Vitamins are often taken in large doses, from 10 to 1000 times the daily requirement.

Large quantities of water-soluble vitamins seldom cause problems because the excess is excreted in the urine. A dietary excess of fat-soluble vitamins is much more likely to cause trouble. When the dietary supply is excessive, the tissue lipids absorb the additional vitamins. Because these fat-soluble vitamins will later diffuse back into circulation, once the symptoms of hypervitaminosis appear they are likely to persist.

When absorbed in massive amounts, fat-soluble vitamins can produce acute symptoms of *vitamin toxicity*. Vitamin A toxicity is the most common condition; it sometimes afflicts children whose parents are overanxious about proper nutrition and vitamins. A single enormous overdose can produce nausea, vomiting, headache, dizziness, lethargy, and even death. Chronic overdose can lead to hair loss, joint pain, hypertension, weight loss, and liver enlargement.

Earlier in this century, vitamin A toxicity was a problem for mariners and Arctic explorers. Several marine animals, including some species of sea turtles, many sharks, and all polar bears, store large quantities of vitamin A in their livers. Unwary explorers who dined on polar bear livers and hungry sailors who feasted on shark or turtle livers sometimes developed chronic or acute symptoms of vitamin A toxicity.

NEWS

Alcohol: A Risky Diversion   *Page 850*

Alcohol production and sales are big business throughout the Western world. Beer commercials on television, billboards advertising various brands of liquors, characters on screen enjoying a drink—all demonstrate the significance of alcohol in our society. Most people are unaware of the medical consequences of this cultural fondness for alcohol. Problems with alcohol are usually divided into those stemming from alcohol abuse and those involving alcoholism. The boundary between these conditions is rather hazy. **Alcohol abuse** is the general term for overuse and the resulting behavioral and physical effects of overindulgence. **Alcoholism** is chronic alcohol abuse with the physiological changes associated with addiction to other CNS-active drugs. Alcoholism has received the most attention in recent years, although alcohol abuse—especially when combined with driving an automobile—is now in the limelight.

Consider the following:

- Alcoholism affects more than 10 million people in the United States alone.

- Alcoholism is probably the most expensive health problem today, with an annual estimated *direct* cost of more than $110 billion. Indirect costs, in terms of damage to automobiles, property, and innocent pedestrians, are unknown.

- An estimated 25–40 percent of U.S. hospital patients are undergoing treatment related to alcohol consumption. There are approximately 200,000 deaths annually due to alcohol-related medical conditions. Some major clinical conditions are caused almost entirely by alcohol consumption. For example, alcohol is responsible for 60–90 percent of all liver disease in the United States.

- Alcohol affects all physiological systems. Major clinical symptoms of alcoholism include: (1) disorientation and confusion (nervous system); (2) ulcers, diar-

121

rhea, and cirrhosis (digestive system); (3) cardiac arrhythmias, cardiomyopathy, anemia (cardiovascular system); (4) depressed sexual drive and testosterone levels (reproductive system); and (5) itching and angiomas (integumentary system).

- The toll on newborn infants has risen steadily over the past 30 years as the number of women drinkers increased. Women consuming 1 ounce of alcohol per day during pregnancy have a higher rate of spontaneous abortion and produce children with lower birth weights. Heavier drinking causes **fetal alcohol syndrome** (FAS). This condition is marked by characteristic facial abnormalities, a small head, slow growth, and mental retardation.

- Perhaps most disturbing of all, the problem of alcohol abuse is considerably more widespread than alcoholism. Although the medical effects are less well documented, they are certainly significant.

Several factors interact to produce alcoholism. The primary risk factors are gender (males are more likely to become alcoholics than females) and a family history of alcoholism. There does appear to be a genetic component, but the relative importance of genes versus social environment has been difficult to assess. It is likely that alcohol abuse and alcoholism can result from a variety of factors.

Treatment may consist of counseling and behavior modification. To be successful, treatment must involve a total avoidance of alcohol. Supporting groups, such as Alcoholics Anonymous, can be very helpful in providing a social framework for abstinence. Use of the drug *disulfiram* (*Antabuse*®) has not proven to be as successful as originally anticipated. Antabuse sensitizes the individual to alcohol so that a drink produces intense nausea, and it was anticipated that this would be an effective deterrent. Clinical tests indicated that it could increase the time between drinks, but not prevent drinking altogether.

 Nutrition and Nutritionists *Page 850*

Professional nutritionists are trained to modify the daily recommendations for specific individuals. There are many complexities involved in determining individual requirements, and a good nutritionist must be part biochemist, part pharmacist, and part clinician. He or she must also be able to work with people who have strong biases about what is acceptable or unacceptable in their diet—opinions that often stem from learned, social, or traditional values that have little correlation with nutritional facts. In other words, good nutritionists are highly skilled, highly motivated individuals who have endured a long educational process. But unlike the situation with most other medically related professions, there are as yet no national standards determining the use of the name nutritionist or qualifications for establishing a practice. You have doubtless encountered self-taught and self-confident "nutritionists" working in health-food stores or supermarkets or selling various products door-to-door. The differences between a balanced diet and one that produces avitaminosis, hypervitaminosis, or a degree of malnutrition are very slim indeed, so if you are seeking nutritional advice, find a qualified professional nutritionist!

Hypothermia

Page 855

The function of the heat-gain center of the brain is to prevent **hypothermia** (hī-pō-THER-mē-a), or below-normal body temperature. If body temperature continues to decline, the thermoregulatory system begins to lose sensitivity and effectiveness. Cardiac output falls, respiratory rate decreases, and if the core temperature falls below 28° C (82°F) cardiac arrest is likely. The individual then has no heartbeat, no respiratory rate, and no response to external stimuli, even painful ones. Body temperature continues to decline, and the skin turns blue and cold.

At this point we would probably assume that the individual has died. But because of the systemwide decrease in metabolic activities, the victim may still be saved, even after several hours have elapsed. Treatment consists of cardiopulmonary support and gradual rewarming both external and internal. The skin can be warmed up to 45° C (110° F) without damage; warm baths or blankets can be used. One effective method of raising internal temperatures involves the introduction of warm saline into the peritoneal cavity.

Hypothermia is a significant risk for those engaged in water sports, and its presence may complicate treatment of a drowning victim. Water absorbs heat roughly 27 times faster than air, and the heat-gain mechanisms are unable to keep pace over long periods or when faced with a large temperature gradient. But hypothermia in cold water does have a positive side. On several occasions small children who have drowned in cold water have been successfully revived after periods of up to 4 hours. Children lose body heat quickly, and their systems stop functioning very quickly as body temperature declines. This rapid drop in temperature prevents the oxygen-starvation and tissue damage that would otherwise occur when breathing stops.

Hypothermia can also be produced intentionally during surgery. Several clinical procedures are possible only because the metabolic rate of a particular organ, or of the entire body, has been decreased drastically. In controlled hypothermia, the individual is first anesthetized to prevent the shivering that would otherwise fight the process.

During open-heart surgery the body is often cooled to 25°–32° C (79°–89° F). This cooling reduces the metabolic demands of the body, which will be receiving blood

123

from an external pump/oxygenator. The heart must be stopped completely during the operation, and it cannot be well supplied with blood over this period. So the heart is exposed to an "arresting solution" at 0°–4° C (32°–39° F) and maintained at a temperature below 15° C (60° F) for the duration of the operation. At these temperatures the cardiac muscle can tolerate several hours of ischemia without damage.

When cardiac surgery is performed on infants, a deep hypothermia may be produced by cooling the entire body to temperatures as low as 11° C (52° F) for an hour or more. In effect this duplicates the conditions experienced by the accidental drowning victims discussed above.

The highly controversial practice of freezing the recently deceased like a TV dinner cannot properly be called hypothermia. Very small organisms can be flash-frozen and subsequently thawed without ill effects because the process is so rapid that ice crystals never form. It will be impossible to flash-freeze anything as large as a human being any time in the foreseeable future. Water expands roughly 7 percent during ordinary freezing, and cell membranes throughout the body are destroyed in the process. Because the damage is done during freezing, advances in thawing technology over the next century will not solve the basic problem.

CHAPTER 26

The Urinary System

Glomerulonephritis

Page 864

Glomerulonephritis (glo-mer-ū-lō-nef-RĪ-tis) is an inflammation of the renal cortex that affects the filtration mechanism. This condition, which may develop after a bacterial infection involving *Streptococcus* bacteria, is an example of an *immune complex disorder*, a class of diseases introduced in Chapter 22 of this Manual. As is often the case with these disorders, the primary infection may not occur in or near the kidneys. However, as the immune system responds to the infection, the number of circulating antigen-antibody complexes skyrockets. These complexes are small enough to pass through the lamina densa, but too large to fit between the slit pores. As a result, the filtration mechanism clogs up, and filtrate production declines. Any condition that leads to such a massive immune response can cause glomerulonephritis, including viral infections and autoimmune disorders.

124

⚕ Polycystic Kidney Disease

Page 867

Polycystic (po-lē-SIS-tik) **kidney disease** is an inherited condition affecting tubular structure. Swellings develop along the length of the tubules, some growing large enough to compress adjacent nephrons and vessels. Kidney function deteriorates, and the kidneys may eventually become nonfunctional. However, the process is so gradual that serious problems seldom appear before the individual is 30–40 years of age. Common symptoms include sharp pain in the sides, recurrent urinary infections, and the presence of blood in the urine. Treatment is symptomatic, focusing on prevention of infection and reduction of pain with analgesics. In severe cases, hemodialysis or kidney transplantation may be required (see Health News: Advances in the Treatment of Renal Failure on pp. 893–894 of the text).

PAH and Analysis of Renal Blood Flow   *Page 868*

Para-aminohippuric acid, or **PAH**, is administered to determine the rate of blood flow through the kidneys. PAH enters the filtrate through filtration at the glomerulus. As blood flows through the peritubular capillaries, any remaining PAH diffuses into the peritubular fluid, and the tubular cells actively secrete it into the filtrate. By the time blood leaves the kidney, virtually all of the PAH has been removed from the circulation and filtered or secreted into the urine. It is therefore possible to calculate renal blood flow if you know the PAH concentrations of the arterial plasma and urine. The calculation proceeds in a series of steps. The first step is to determine the plasma flow through the kidney using the formula:

$$P_f = PAH_u \times \left(\frac{V_u}{PAH_p} \right)$$

P_f is the plasma flow, also known as the *effective renal plasma flow*, or *ERPF*

PAH_u is the concentration of PAH in the urine, usually expressed in terms of milligrams per milliliter (mg/mℓ)

V_u is the volume of urine produced, usually in terms of milliliters per minute

PAH_p is the concentration of PAH in arterial plasma in milligrams per milliliter

Consider the following example: A patient producing urine at a rate of 1 mℓ per minute has a urinary PAH concentration of 15 mg/mℓ with an arterial PAH concentration of 0.02 mg/mℓ. The patient's hematocrit is normal (Hct = 45).

$$P_f = \frac{15 \text{ mg/m}\ell \times 1 \text{ m}\ell/\text{min}}{0.02 \text{ mg/m}\ell} = 750 \text{ m}\ell/\text{min}$$

This value is an estimate of the plasma flow through the glomeruli and around the kidney tubules each minute. But as you will recall from Chapter 19, plasma accounts for only part of the volume of whole blood. The rest consists of blood cells. So to have a plasma flow of 750 mℓ/min, the blood flow must be considerably greater. The patient's hematocrit is 45, which means that plasma accounts for 55 percent of the whole blood volume. So to calculate the renal blood flow we must multiply the plasma volume by 1.8 (each 100 mℓ of blood has 55 mℓ of plasma, and 100/55 = 1.8):

$$750 \text{ m}\ell/\text{min} \times 1.8 = 1350 \text{ m}\ell/\text{min}$$

This value, 1350 mℓ/min, is the estimated tubular blood flow. The final step is to adjust this figure to account for blood that enters the kidney but flows to the renal pelvis, the capsule, or other areas not involved with urine production. This value is usually estimated as 10 percent of the total blood flow. To complete the calculation for this example:

if 1350 mℓ/min = 90 percent of total blood flow
then 10 percent = 1350/9 = 150 mℓ/min
and total blood flow = 1350 + 150 = 1500 mℓ/min.

Estimating the GFR

Page 875

A **creatinine clearance test** is often used to estimate the GFR. Creatinine results from the breakdown of creatine in muscle tissue, and it is normally excreted in the urine. Creatinine enters the filtrate at the glomerulus, and it is not reabsorbed in significant amounts.

If the amount in the urine and the concentration in the plasma are known, the filtration rate can be estimated. For example, consider a patient who excretes 84 mg of creatinine each hour with a plasma creatinine concentration of 1.4 mg/dℓ. The filtration rate is equal to the amount secreted divided by the plasma concentration: (83 mg/hr) ÷ (1.4 mg/dℓ) = 60 dℓ/hr, or 6000 mℓ per hour. GFR is usually reported in terms of milliliters per minute, and 6000 mℓ per hour is equal to 100 mℓ per minute (6000/60 = 100).

This figure is only an approximation of the GFR, because a small and variable amount of creatinine enters the urine (up to 15 percent) through active tubular secretion. However, the creatinine clearance test has advantages because no special procedures are involved; creatinine levels are routinely measured in clinical blood and urine tests. When necessary, a more accurate GFR determination can be performed using a complex carbohydrate, *inulin*, that is not metabolized in the body and is neither reabsorbed nor secreted by the kidneys.

Conditions Affecting the Filtration Process
Page 875

Abnormal changes in the pressures that determine filtration pressure (P_f) can result in significant alterations in kidney function. Examples of factors that can disrupt normal filtration rates include *physical damage to the filtration apparatus* and *interference with normal filtrate or urine flow*.

Physical damage to the filtration apparatus   The lamina densa and podocytes can be injured by mechanical trauma, such as a blow to the kidneys, bacterial infection, or exposure to metabolic poisons, such as mercury. The usual result is a sudden increase in the permeability of the glomerulus. When damage is severe, plasma proteins and even blood cells enter the capsular spaces. The loss of plasma proteins has two immediate effects: (1) it reduces the osmotic pressure of the blood, and (2) it increases the osmotic pressure of the filtrate. The result is an increase in the net filtration pressure and an increased rate of filtrate production.

Blood cells entering the filtrate will not be reabsorbed. The presence of blood cells in the urine is called **hematuria** (hēm-a-TŪR-ē-a). Although small amounts of protein can be reabsorbed, when glomeruli are severely damaged the nephrons are unable to reabsorb all of the plasma proteins entering the filtrate. Plasma proteins then appear in the urine, a condition termed **proteinuria** (prō-tēn-ŪR-ē-a). Proteinuria and hematuria indicate that kidney damage has occurred.

Interference with filtrate or urine flow   If the tubule, collecting duct, or ureter becomes blocked and urine flow cannot occur, capsular pressures gradually rise. When the capsular hydrostatic pressure and blood osmotic pressure equal the glomerular hydrostatic pressure, filtration stops completely. The severity of the problem depends upon the site of the blockage. If it involves a single nephron, only a single glomerulus will be affected. If the blockage occurs within the ureter, filtration in that kidney will come to a halt. If the blockage occurs in the urethra, both kidneys will become nonfunctional. (Examples of factors involved in urinary blockage are discussed below—see Problems with the Conducting System.)

Elevated capsular pressures can also result from inflammation of the kidneys, a condition called **nephritis** (nef-RĪ-tis). A generalized nephritis may result from bacterial infections or exposure to toxic or irritating drugs. One of the major problems in nephritis is that the inflammation causes swelling, but the renal capsule prevents

127

an increase in the size of the kidney. The result is an increase in the hydrostatic pressures in the peritubular fluid and filtrate. This pressure opposes the glomerular hydrostatic pressure, lowering the net filtration pressure and the GFR.

Inherited Problems with Tubular Function

Page 876

The tubular absorption of specific ions or compounds involves many different carrier proteins. Some individuals have an inherited inability to manufacture one or more of these carrier proteins, and so experience impaired tubular function. For example, in **renal glycosuria** (glī-cō-SŪ-rē-a) a defective carrier protein makes it impossible for the PCT to reabsorb glucose from the filtrate. Although renal glucose levels are abnormally high, blood glucose is normal, which distinguishes this condition from diabetes mellitus. Affected individuals generally do not have any clinical problems except when demand for glucose is high, as in starvation, acute stress, or pregnancy.

There are several types of **aminoaciduria** (a-mē-nō-as-i-DŪ-rē-a), differing according to the identity of the missing carrier protein. Some of these disorders affect reabsorption of an entire class of amino acids; others involve individual amino acids, such as lysine or histidine. **Cystinuria** is the most common disorder of amino acid transport, occurring in 1 person in 10–15,000. Persons with this condition have difficulty reabsorbing cystine and amino acids with similar carbon structure, such as lysine, arginine, and ornithine. The most obvious and painful symptom is the formation of kidney and bladder stones containing crystals of these amino acids. In addition to removal of these stones, treatment for cystinuria involves maintaining a high rate of urinary flow, so that amino acid concentrations do not rise high enough to promote stone formation, and reducing urinary acidity, since stone formation is enhanced by acid conditions.

Any of these problems with tubular absorption and secretion will have a direct effect on urinary concentration. The greater the number of solutes in the filtrate, the less water can be extracted by osmosis.

Diuretics

Page 883

Diuretics (dī-ū-RET-iks) are drugs administered to promote the loss of water in the urine. The usual goal is the reduction of blood volume, blood pressure, or both. The ability to control renal water losses with relatively safe and effective diuretics has been a lifesaver for many patients, especially those with high blood pressure or congestive heart failure.

Diuretics have many different mechanisms of action, but all affect transport activities and/or water reabsorption along the nephron and collecting system. Important diuretics in use today include:

Osmotic diuretics: **Osmotic diuretics** are metabolically harmless substances that are filtered at the glomerulus and ignored by the tubular epithelium. Their presence in the urine increases its osmolarity and limits the amount of water reabsorption possible. **Mannitol** (MAN-i-tol) is the most frequently administered osmotic diuretic. It is used to accelerate fluid loss and speed the removal of toxins from the blood, and to elevate the GFR after severe trauma or other conditions have impaired renal function.

Drugs that block sodium and chloride transport: A class of drugs called **thiazides** (THĪ-a-zīdz) reduce sodium and chloride transport in the proximal and distal tubules. Thiazides such as *chlorothiazide* are often used to accelerate fluid losses in the treatment of hypertension and peripheral edema. One sometimes undesirable effect of these diuretics is increased potassium loss in the urine.

High ceiling diuretics: The **high ceiling diuretics**, such as *furosemide* and *bumetanide*, inhibit transport along the loop of Henle, reducing the osmotic gradient and the ability to concentrate the urine. They are called "high ceiling" diuretics because they produce a much higher degree of diuresis than other drugs, with a corresponding increase in the rate of urinary potassium loss. They are fast-acting and are often used in a clinical crisis—for example, in treating acute pulmonary edema.

Aldosterone blocking agents: Blocking the action of aldosterone prevents the reabsorption of sodium along the DCT and collecting tubule, and so accelerates fluid losses. The drug *spironolactone* is an example of this type of diuretic. It is often used in conjunction with other diuretics because blocking the aldosterone-activated exchange pumps helps reduce the potassium ion loss. These drugs are also known as *potassium sparing diuretics*.

Drugs with diuretic side effects: Many drugs prescribed for other conditions promote diuresis as a side effect. For example, drugs that block carbonic anhydrase activity, such as *acetazolamide* (*Diamox*), have an indirect effect on sodium transport (see Figure 26-13 on p. 877 of the text). Although they cause diuresis, these drugs are seldom prescribed with that in mind. (Because carbonic anhydrase is also involved in aqueous human secretion, Diamox is used to reduce intraocular pressure in glaucoma patients—see Chapter 17 of this Manual.) Two more familiar drugs, caffeine and alcohol, have pronounced diuretic effects. Caffeine produces diuresis directly, by reducing sodium reabsorption along the tubules. Alcohol works indirectly, by suppressing the release of ADH by the posterior pituitary gland.

Average Values for Urinalysis

Page 885

The Diagnostics box on p. 885 of the text discusses the techniques of urinalysis and their clinical usefulness. For reference, Table 26-A presents typical values for the most important components of normal urine.

Problems with the Conducting System

Page 890

Local blockages of the collecting tubules, collecting ducts, or ureter may result from the formation of **casts**, small blood clots, epithelial cells, lipids, or other materials. Casts are often excreted in the urine and visible in microscopic analysis of urine samples. **Calculi** (KAL-kū-lī), or "kidney stones," form from calcium deposits, magnesium salts, or crystals of uric acid. This condition is called **nephrolithiasis** (nef-rō-li-THĪ-a-sis). The blockage of the urinary passage by a stone or other factors, such as external compression, results in **urinary obstruction**.

Obstruction of the ureter usually leads to **pyelitis**, an inflammation of the renal pelvis. Symptoms include fever, pain on the affected side, nausea, and a reduction in urine volume. Kidney stones are usually visible on an X-ray, and if peristalsis and fluid pressures are insufficient to dislodge them they must be surgically removed or destroyed. One interesting nonsurgical procedure involves immersing the patient in water and then blasting the kidney stone apart with focused sound waves. The apparatus used is called a **lithotripter**. Another entails the insertion of a catheter armed with a laser that can shatter the blockage with intense light beams.

One of the more bizarre causes of urinary obstruction involves a small catfish, the Candiru (*Vandellia cirrhosa*). This slender inhabitant of the Amazon River normally lives a parasitic existence, wriggling its way into the gill cavities of larger fishes. Candiru have been known to enter the urethrae of bathers, where they promptly become trapped. Spines on the head and fins prevent its retreat, and surgery must be performed to prevent a fatal urinary obstruction.

Problems with the Micturition Reflex

Page 890

Incontinence (in-KON-ti-nens) refers to an inability to voluntarily control urination. This may reflect damage to the CNS, the spinal cord, or the nerve supply to the bladder or external sphincter. Incontinence often accompanies Alzheimer's disease,

■ TABLE 26-A Typical values from standard urine testing

Compound	Primary Source	Daily Excretion[a]	Concentration	Remarks
Nitrogenous Wastes				
Urea	Deamination of amino acids at liver and kidneys	21g	1.8 g/dℓ	Rises if negative nitrogen balance exists
Creatinine	Breakdown of creatine phosphate in skeletal muscle	1.8 g	150 mg /dℓ	Proportional to muscle mass; decreases during atrophy or muscle disease
Ammonia	Deamination by liver and kidney, absorption from intestinal tract	0.68 g	60 mg/dℓ	
Uric acid	Breakdown of purines	0.53 g	40 mg/dℓ	Increases in gout, liver diseases
Hippuric acid	Breakdown of dietary toxins	4.2 mg	350 µg/dℓ	
Urobilinogen	Asborption at colon	1.5 mg	125 µg/dℓ	Gives urine its yellow color
Bilirubin	Hemoglobin breakdown product, released by RES	0.3 mg	20 µg/dℓ	Increase may indicate problem with liver excretion or excess production; causes yellow skin color in jaundice
Nutrients and Metabolites				
Carbohydrates		0.11 g	9 µg/dℓ	Primarily glucose; glycosuria if T_m exceeded
Ketone bodies		0.21 g	17 µg/dℓ	Ketonuria may occur during postabsorptive state
Lipids		0.02 g	1.6 µg/dℓ	May increase in some kidney diseases
Amino acids		2.25 g	287.5 µg/dℓ	Note relatively high loss compared to other metabolites—low T_m, no large storage pools exist; excess (aminoaciduria) indicates T_m problem
Ions				
Sodium		4.0 g	333 mg/dℓ	Varies with diet, urine pH, hormones, etc.
Chloride		6.4 g	533 mg/dℓ	
Potassium		2.0 g	166 mg/dℓ	Varies with diet, urine pH, hormones, etc.
Calcium		0.2 g	17 mg/dℓ	Hormonally regulated (PTH/CT)
Magnesium		0.15 g	13 mg/dℓ	
Blood cells[b]				
RBCs		130,000/day	100/mℓ	Excess (hematuria) indicates vascular damage
WBCs		650,000/day	500/mℓ	Excess (pyuria) indicates infection, inflammation

[a] Representative values for a 70-kg male.
[b] These are usually estimated by counting the numbers in a sample of sediment after urine centrifugation.

and it may also result from a stroke or spinal cord injury. In most cases the individual develops an **automatic bladder**. The micturition reflex remains intact, but voluntary control of the external sphincter is lost and the individual cannot prevent the reflexive emptying of the bladder.

Paraplegic or quadriplegic patients usually develop an automatic bladder once the initial period of spinal shock has ended. Over the interim the urethra must be catheterized to permit normal urinary drainage. Damage to the pelvic nerves can eliminate the micturition reflex entirely, because these nerves carry both afferent and efferent fibers of this reflex arc. In this case the bladder becomes greatly distended with urine. It remains filled to capacity, and the excess trickles into the urethra in an uncontrolled stream.

Infants lack voluntary control over urination because the necessary corticospinal connections have yet to be established. Toilet training before age 2 usually involves training the parent to anticipate the timing of the reflex, rather than training the child to exert conscious control. Childbirth can stretch and damage the sphincter muscles, and some women then develop *stress incontinence*. In this condition elevated intraabdominal pressures, such as during a cough or sneeze, can overwhelm the sphincter muscles, causing urine to leak out.

CHAPTER 27

Fluid, Electrolyte, and Acid-Base Balance

 Water Excess and Depletion *Page 900*

The body's water content cannot easily be determined. However, the concentration of sodium, the most abundant ion in the ECF, provides useful clues to the state of water balance. When the body water content rises, the sodium concentration of the extracellular fluid falls below 130 mEq/ℓ. This condition is termed **hyponatremia** (*natrium*, sodium). When the body water content declines, the sodium concentration rises; when it exceeds 150 mEq/ℓ, **hypernatremia** exists.

Hyponatremia is a sign of **overhydration**, or **water excess**. A temporary, acute hyponatremia may be caused by the infusion of hypotonic fluids or the ingestion of massive amounts of fresh water. The osmotic changes, especially the reduction in sodium ion concentrations, affect CNS function. In the early stages of hyponatremia the individual behaves as if drunk on alcohol. This condition, called **water intoxication**, sounds funny, but it is extremely dangerous. Untreated cases can rapidly progress from confusion to hallucinations, convulsions, coma, and death. Treatment of severe water intoxication usually involves diuretics and infusing a concentrated salt solution that elevates sodium levels to near normal.

An inability to control osmolarity by eliminating excess water can also cause hyponatremia. Several conditions discussed elsewhere such as chronic renal failure (p. 886 of the text), congestive heart failure (Chapter 21 of this Manual), and cirrhosis (Chapter 24 of this Manual), may promote hyponatremia and an elevated ECF volume. Peripheral edema, CNS disturbances similar to those described for water intoxication, and muscle spasms may develop if the problem is not corrected.

Hyponatremia can also result from endocrine disorders. One important endocrine condition is the **syndrome of inappropriate ADH secretion (SIADH)** (see Table 18-7 on p. 602 of the text). In SIADH, ADH release occurs at high levels regardless of the osmolarity of the plasma. Water is conserved, and the ECF volume gradually enlarges. As the blood volume increases, so does the rate of glomerular filtration. Eventually the rate of filtrate production exceeds the reabsorption capabilities of the tubules, despite the high ADH levels, and urine production accelerates. Along with water, the kidneys lose an abnormally large number of sodium ions in the urine, and this leads to hyponatremia.

Treatment for all forms of hyponatremia involve (1) restricting water intake and (2) accelerating water loss through diuretics. (The infusion of a hypertonic electrolyte solution is an emergency measure, taken to control severe symptoms of water intoxication.) When hyponatremia results from a disorder such as kidney failure or congestive heart failure, the underlying problem must also be corrected.

Dehydration, or **water depletion**, develops when water losses outpace water gains. Plasma osmolarity gradually increases, and hypernatremia results. The loss of body water is associated with severe thirst, a dryness and wrinkling of the skin, and a fall in plasma volume and blood pressure. Eventually, circulatory shock develops, usually with fatal consequences.

Treatment for dehydration entails administering hypotonic fluids by mouth or via intravenous infusion. This increases ECF volume and restores normal electrolyte concentrations.

Hyperkalemia and Hypokalemia *Page 904*

When the plasma concentration of potassium falls below 2 mEq/ℓ, extensive muscular weakness develops, followed by eventual paralysis. This condition, called **hypo-**

kalemia (*kalium*, potassium), in connection with ion effects on cardiac function. (See p. 651 of the text and the section on Cardiac Contractions and the Composition of the Extracellular Fluid in Chapter 20 of this Manual.) Causes of hypokalemia include:

1. **Inadequate dietary intake:** If potassium gains from the diet do not keep pace with the rate of potassium loss in the urine, potassium concentrations in the ECF will decline.

2. **Administration of diuretic drugs:** Several diuretics, including *Lasix*, can produce hypokalemia by increasing the volume of urine produced. Although the concentration of potassium in the urine is low, the greater the total volume, the larger the amount of potassium lost.

3. **Excessive aldosterone secretion:** The condition of *aldosteronism*, characterized by excessive aldosterone secretion was discussed in Chapter 18 of this Manual (see Disorders of the Adrenal Cortex). This condition results in hypokalemia because the reabsorption of sodium ions is tied to the secretion of potassium.

4. **An increase in the pH of the extracellular fluids:** When the hydrogen ion concentration declines in the ECF, cells exchange intracellular hydrogen ions for extracellular potassium. This ion swap helps stabilize the extracellular pH, but gradually lowers the potassium ion concentration in the ECF.

Treatment for hypokalemia usually includes (1) increasing dietary intake, by salting food with potassium salts (KCl) or by taking potassium tablets, such as *Slow-K*, and in severe cases (2) the infusion of a solution containing potassium ions at a concentration of 40–60 mEq/ℓ.

High potassium ion concentrations in the ECF produce a condition known as **hyperkalemia** that can be even more dangerous than hypokalemia. Severe cardiac arrhythmias appear when the potassium ion concentration exceeds 8 mEq/ℓ; the mechanism was discussed in Chapter 20 of this Manual (see Cardiac Contractions and the Composition of the Extracellular Fluid). Hyperkalemia may result from:

1. **Renal failure:** Kidney failure due to damage or disease will prevent normal potassium ion secretion and so produce hyperkalemia.

2. **The administration of diuretic drugs that block sodium reabsorption:** Potassium secretion is linked to sodium reabsorption. When sodium reabsorption slows down, so does potassium secretion, and hyperkalemia can result.

3. **A decline in the pH of the ECF:** When the pH in the ECF declines, hydrogen ions move into the ICF in exchange for intracellular potassium ions. In addition, at the kidney tubules potassium secretion slows down because hydrogen ions are secreted instead of potassium ions. The combination of increased potassium entry into the ECF and decreased potassium secretion can produce a dangerous hyperkalemia very rapidly.

134

Treatment for hyperkalemia includes (1) elevation of ECF volume with a solution low in potassium, (2) stimulation of urinary potassium loss using appropriate diuretics, such as *Lasix*, (3) administration of buffers (usually sodium bicarbonate) that can control the pH of the ECF, (4) restriction of dietary potassium intake, and (5) administration of enemas or laxatives containing compounds, such as *kayexolate*, that promote potassium loss across the digestive lining. In cases resulting from renal failure, kidney dialysis may be required as well.

Classifying Acid-Base Disorders   *Page 915*

Virtually every patient with a problem affecting the cardiovascular, respiratory, urinary, digestive, or nervous systems may have potentially dangerous problems with acid-base balance. For this reason blood tests used in patient evaluation include tests designed to provide information on pH and buffer function. Standard tests include blood pH, pCO_2, and bicarbonate levels. Using these data, the condition can usually be diagnosed relatively easily. Table 27-2 on p. 915 of the text indicates the patterns that characterize the four major categories of acid-base disorders. It is often useful to employ a graphical representation of the data on bicarbonate, carbon dioxide, pH, and pCO_2 values. This graph, called a **nomogram**, is shown in Figure 27-A. The horizontal axis represents pH and the vertical axis represents plasma HCO_3^- concentration. The curving lines indicate the relationship between pH and bicarbonate levels at a specific pCO_2. For example, at a pCO_2 of 30 mm Hg (the curve that starts in the upper right-hand corner of the nomogram) the pH and bicarbonate values must lie somewhere along that line.

The shaded area at the center of the graph corresponds to the normal range of pH, bicarbonate, and pCO_2 levels. When acid-base disorders occur, these values change. The changes, when plotted on the nomogram, will roughly follow one of the heavy arrows, each of which characterizes a particular clinical condition.

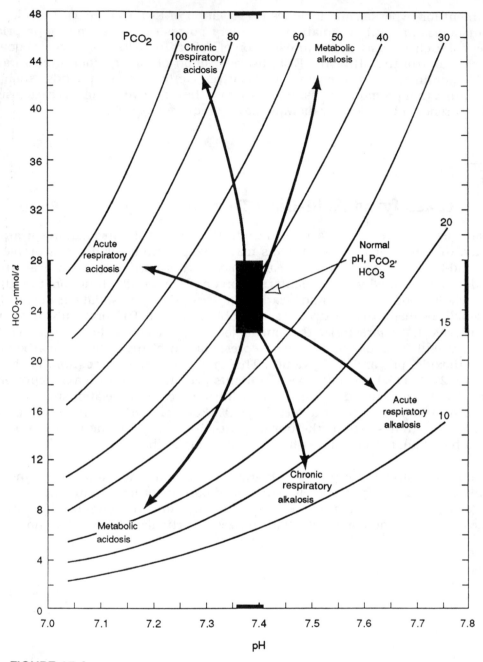

FIGURE 27-A
Characteristics of acid-base disorders

CHAPTER 28

The Reproductive System

Cryptorchidism

Page 919

In **cryptorchidism** (krip-TŌR-ki-dizm) the testes have not descended into the scrotum by the time of birth. This condition occurs in about 3 percent of full-term deliveries and in roughly 30 percent of premature births. In most instances normal descent occurs a few weeks later, but the condition can be surgically corrected if it persists. Corrective measures are usually taken before puberty because cryptorchid (abdominal) testes will not produce sperm, and the individual will be sterile. If the testes cannot be moved into the scrotum, they will usually be removed, because about 10 percent of those with uncorrected cryptorchid testes eventually develop testicular cancer.

Testicular Torsion

Page 923

Because the testes are relatively loosely attached to the scrotal walls, they may become twisted within the scrotal cavity. This condition, called **testicular torsion**, usually occurs in children or adolescents. Symptoms include pain in the groin and inguinal region, local inflammation, and swelling of the scrotum on the affected side. Treatment involves prompt external or surgical manipulation of the testis to relieve the twisting. Because any kinks in the spermatic cord severely restrict the arterial supply to the testis, corrective measures must be taken within 4–6 hours, before the testicular tissues become permanently damaged. If deprived of circulation for longer periods, the damage will be irreversible, and the affected testis will have to be removed. This surgical procedure is called an **orchiectomy** (ōr-kid-EK-to-mē; *orchis*, testis); the common term *castration* indicates that both testes have been removed.

Prostatic inflammation, or **prostatitis** (pros-ta-TĪ-tis), can occur at any age, but it most often afflicts older men. Prostatitis may result from bacterial infections, but the condition may also develop in the apparent absence of pathogens. Individuals with prostatitis complain of pain in the lower back, perineum, or rectum, sometimes accompanied by painful urination and the discharge of mucus secretions from the urethral meatus. Antibiotic therapy is usually effective in treating cases resulting from bacterial infection, but in other cases antibiotics may not provide relief.

Prostatic enlargement usually occurs spontaneously in men over age 50. The increase in size occurs at the same time that hormonal changes are under way within the testes. Androgen production by the interstitial cells decreases over this period, and at the same time these endocrine cells begin releasing small quantities of female sex hormones, called *estrogens*, into the circulation. The combination of lower testosterone levels and the presence of estrogen probably stimulates the prostatic growth. In severe cases, prostatic swelling can constrict and block the urethra and even the rectum. The urinary obstruction can cause permanent kidney damage if not corrected. Partial surgical removal is the most effective treatment at present. In the procedure known as a **TURP** (transurethral prostatectomy), an instrument pushed along the urethra restores normal function by cutting away the swollen prostatic tissue. Most of the prostate remains in place, and there are no external scars.

Prostate cancer is the second most common cancer in men, and it is the second most common cause of cancer deaths in males. Each year 103,000 cases are diagnosed in the United States, and there are approximately 28,500 deaths. Most patients are elderly (over age 65). There are racial differences in susceptibility that are poorly understood; the incidence is relatively high among American blacks and low among Asians.

Prostate cancer usually originates in one of the secretory glands, and as it progresses it produces a nodular lump or swelling on the prostatic surface. Palpation of the prostate gland through the rectal wall is the easiest diagnostic screening procedure, but transrectal prostatic ultrasound provides more detailed information.

If the condition is detected before the cancer cells have spread to other organs, the usual treatment is either localized radiation or the surgical removal of the prostate gland. This operation, called a **prostatectomy** (pros-ta-TEK-to-mē), is often effective in controlling the condition, but undesirable side effects may include a loss of sexual function and urinary incontinence. Modified surgical procedures can reduce these risks and maintain normal sexual function in almost 3 out of 4 patients.

One recent screening method involves a blood test for *prostate-specific antigen*. Elevated levels of this antigen, normally present in low concentrations, may indicate the presence of prostate cancer. This test is more sensitive than the serum enzyme assay previously used for screening purposes. The enzyme test for *prostatic acid phosphatase*, detects prostate cancer in comparatively late stages of development.

Once metastasis is under way, and the lymphatic system, lungs, bone marrow, liver, or adrenal glands are involved, the survival rates are significantly lower. Potential treatments for metastatic prostate cancer include more intensive radiation dosage, hormonal manipulation, lymph node removal, and aggressive chemotherapy. Because the cancer cells are stimulated by testosterone, treatment may involve castration or hormones that depress GnRH or LH production. Until recently the usual hormone selected was *diethylstilbestrol* (DES), an estrogen. Two other options are (1) *drugs that mimic GnRH*: these drugs are given in high doses, producing a surge in LH production followed by a sharp decline to very low levels, presumably as the endocrine cells adapt to the excessive stimulation, and (2) *drugs that block the action of androgens*: several new drugs, including *flutamide*, block the cytoplasmic receptors for testosterone and prevent stimulation of the cancer cells. However, despite these interesting advances in treatment, the average survival time for patients diagnosed with advanced prostatic cancer is only 2.5 years.

Uterine Tumors   *Page 932*

Uterine tumors are the most common tumors in women. It has been estimated that 40 percent of women over age 50 have benign uterine tumors involving smooth muscle and connective tissue cells. These **leiomyomas** (lē-ō-mī-Ō-mas), or *fibroids*, are stimulated by estrogens, and they can grow quite large, reaching weights as great as 13.6 kg (30 lb) if left untreated. Occlusion of the uterine tubes, distortion of adjacent organs, and compression of blood vessels may lead to a variety of complications. In young women, observation or conservative treatment with drugs or restricted surgery may be utilized. In older women, a decision may be made to remove the entire uterus, a procedure termed a **hysterectomy** (his-ter-EK-to-mē).

Benign epithelial tumors in the uterus are called **endometrial polyps**. Roughly 10 percent of women probably have polyps, but because of their small size and lack of symptoms the condition passes unnoticed. If bleeding occurs or if the polyps become excessively enlarged, they can be surgically removed.

Uterine cancers are less common, affecting approximately 11.9 per 100,000 women. In 1990 there were roughly 97,000 new cases reported in the United States and roughly 10,000 deaths. There are two types of uterine cancers, **endometrial** and **cervical**.

Endometrial cancer is an invasive cancer of the endometrial lining. There are roughly 34,000 cases reported each year in the United States, with approximately 3000 deaths. The condition most often affects women age 50–70. Estrogen therapy, used to treat osteoporosis in post-menopausal women (see the discussion in Chapter 7 of the text) increases the risk of endometrial cancer by 2–10 times. Adding cyclic progesterone therapy to the estrogen therapy seems to reduce this risk.

There is no satisfactory screening test for endometrial cancer. The most common symptom is irregular bleeding, and diagnosis typically involves examination of a biopsy of the endometrial lining by suction or scraping. The prognosis varies with the degree of metastasis. Treatment of early-stage endometrial cancer involves a hysterectomy followed by localized radiation therapy. In advanced stages, more aggressive radiation treatment is recommended. Chemotherapy has not proven to be very successful in treating endometrial cancers, and only 30–40 percent of patients benefit from this approach.

Cervical cancer is the most common reproductive system cancer in women age 15–34. Roughly 13,000 new cases of invasive cervical cancer are diagnosed each year in the United States, and approximately 40 percent of them will eventually die of this condition. Another 50,000 patients are diagnosed with a less aggressive form of cervical cancer.

Most women with cervical cancer fail to develop symptoms until late in the disease. At that stage, vaginal bleeding, especially after intercourse, pelvic pain, and vaginal discharge may appear. *Early detection is the key to reducing the mortality rate for cervical cancer.* The standard screening test is the **Pap smear**, named for Dr. George Papanicolaou, an anatomist and cytologist. The cervical epithelium normally sheds its superficial cells, and a sample of cells scraped or brushed from the epithelial surface can be examined for abnormal or cancerous cells. The American Cancer Society recommends yearly Pap tests at ages 20 and 21, followed by smears at 1- to 3-year intervals until age 65.

The primary risk factor for this cancer is a history of multiple sexual partners. It appears likely that these cancers develop following viral infection by one of several different *human papilloma viruses* (HPV) that can be transmitted through sexual contact.

Treatment of localized, noninvasive cervical cancer involves the removal of the affected portion of the uterus. Treatment of more advanced cancers typically involves a combination of radiation therapy, hysterectomy, lymph node removal, and chemotherapy.

140

There are several different forms of vaginitis, and minor cases are relatively common. **Candidiasis** (kan-di-DĪ-a-sis) results from a fungal (yeast) infection. The organism responsible appears to be a normal component of the vaginal ecosystem in 30–80 percent of normal women. Antibiotic administration, immunosuppression, stress, diabetes mellitus, pregnancy, and other factors that change the local environment can stimulate the unrestricted growth of the fungus. Symptoms include itching and burning sensations, and a lumpy white discharge may also be produced. Topical antifungal medications are used to treat this condition. *Monistat* and *Gyne-Lotrimin* are two medicines now available on a non-prescription basis to treat this condition.

Bacterial (*nonspecific*) **vaginitis** results from the combined action of several bacteria. The bacteria involved are normally present in about 30 percent of adult women. In this form of vaginitis the vaginal discharge contains epithelial cells and large numbers of bacteria. The discharge has a homogeneous, sticky texture and a characteristic odor sometimes described as "fishy or amine-like." Antibiotics are often effective in controlling this condition.

Trichomoniasis (trik-ō-mō-NĪ-a-sis) involves infection by a parasite, *Trichomonas vaginalis*, introduced by sexual contact with a carrier. Because it is a sexually transmitted disease, both partners must be treated to prevent reinfection.

A more serious vaginal infection by *Staphylococcus* bacteria is responsible for symptoms of **toxic shock syndrome** (**TSS**). Symptoms include high fever, sore throat, vomiting and diarrhea, and a generalized rash. As the condition progresses, shock, respiratory distress, and kidney or liver failure may develop, and 10–15 percent of all cases prove fatal. These symptoms result from the entry of bacterial toxins and even bacteria into the bloodstream. This disorder was first recognized in 1978, when it affected a group of children. Since that time roughly 2500 cases have been diagnosed, 95 percent of them adult women. Most of these women (over 90 percent) have developed the condition while using a tampon, but the precise link between tampon use and TSS remains uncertain. The use of "superabsorbent" tampons was initially thought to be responsible, and the incidence of TSS did decline as those items were removed from the market. However, TSS continues to occur at a low but significant rate (6.2 per 100,000 menstruating women per year) in those using ordinary tampons and vaginal sponges, and in people of either sex after abrasion or burn injuries that promote bacterial infection. Treatment for TSS involves fluid administration, removal of the focus of infection (such as removal of a tampon or cleansing of a wound), and antibiotic therapy.

Sexually transmitted diseases, or **STDs**, are transferred from individual to individual, usually or exclusively by sexual intercourse. A variety of bacterial, viral, and fungal infections are included in this category. At least two dozen different STDs are currently recognized. All are unpleasant. *Chlamydia*, noted in the Clinical Comment on p. 933 of the text, can cause PID and infertility. Other types of STDs are quite dangerous, and a few, including AIDS, are deadly. Here we will discuss three of the most common sexually transmitted diseases: *gonorrhea*, *syphilis*, and *herpes*.

Gonorrhea: The bacterium *Neisseria gonorrhoeae* is responsible for gonorrhea, one of the most common sexually transmitted diseases in the United States. Nearly 2 million cases are reported each year. These bacteria usually invade epithelial cells lining the male or female reproductive tracts. In relatively rare cases they will also colonize the pharyngeal or rectal epithelium.

The symptoms of genital infection vary, depending on the sex of the individual concerned. It has been estimated that up to 80 percent of women infected with gonorrhea experience no symptoms, or symptoms so minor that medical treatment is thought to be unnecessary. As a result these individuals act as carriers, spreading the infection through their sexual contacts. An estimated 10–15 percent of women infected with gonorrhea experience more acute symptoms because the bacteria invade the epithelia of the uterine tubes. This probably accounts for many of the cases of **pelvic inflammatory disease (PID)** in the U.S. population; as many as 80,000 women may become infertile each year as the result of scar tissue formation along the uterine tubes after gonorrheal infections.

Diagnosis in males seldom poses as great a problem, for all but 20–30 percent of infected males develop symptoms recognized as requiring immediate medical attention. The urethral invasion is accompanied by pain on urination (*dysuria*) and often a viscous urethral discharge. A sample of the discharge can be cultured to permit positive identification of the organism involved. Treatment of gonorrhea involves the administration of antibiotics.

Syphilis: Syphilis (SIF-i-lis) results from infection by the bacterium *Treponema pallidum*. The first reported syphilis epidemics occurred in Europe during the sixteenth century. The death rate from the "great pox" was appalling, far greater than today even after taking into account the absence of antibiotic therapies at that time. It appears likely that the syphilis organism has changed over the interim. These changes have reduced the mortality rate but prolonged the period of illness and increased the likelihood of successful transmission. Despite these relative improvements, syphilis still remains a life-threatening disease. Untreated syphilis can cause serious cardiovascular and neurologic illness years after infection, or

142

it can be spread to the fetus during pregnancy to produce congenital malformations. The annual reported incidence of this disease has now risen to roughly 15 cases per 100,000 population, the highest rate in 40 years. An equivalent or greater number probably went unrecognized or unreported.

Primary syphilis begins as the bacteria cross the mucous epithelium and enter the lymphatics and bloodstream. At the invasion site the bacteria multiply, and after an incubation period ranging from 1.5–6 weeks their activities produce a raised lesion, or **chancre** (SHANG-ker). This lesion remains for several weeks before fading away, even without treatment. In heterosexual men the chancre usually appears on the penis; in women it may develop on the labia, vagina, or cervix. Lymph nodes in the region usually enlarge and remain swollen even after the chancre has disappeared.

Symptoms of **secondary syphilis** appear roughly 6 weeks later. Secondary syphilis usually involves a diffuse, reddish skin rash. Like the chancre, the rash fades over a period of 2–6 weeks. These symptoms may be accompanied by fever, head-aches, and uneasiness. The combination is so vague that the disease may easily be overlooked or diagnosed as something else entirely. In a few instances more serious complications such as meningitis, hepatitis, or arthritis may develop.

The individual then enters the **latent phase**. The duration of the latent phase varies widely. Fifty to seventy percent of untreated individuals with latent syphilis fail to develop the symptoms of **tertiary syphilis**, or *late syphilis*, although the bacterial pathogens remain within their tissues. Those destined to develop tertiary syphilis may do so 10 or more years after infection.

The most severe symptoms of tertiary syphilis involve the CNS and the cardiovascular system. **Neurosyphilis** may result from bacterial infection of the meninges or the tissues of the brain and/or spinal cord. **Tabes dorsalis** (TĀ-bēz dor-SAL-is) results from the invasion and demyelination of the posterior columns of the spinal cord and the sensory ganglia and nerves. In the cardiovascular system the disease affects the major vessels, leading to aortic stenosis, aneurysms, or calcification.

Equally disturbing are the effects of transmission from mother to fetus across the placenta. These cases of **congenital syphilis** are marked by infections of the developing bones and cartilages of the skeleton and progressive damage to the spleen, liver, bone marrow, and kidneys. The risk of transmission may be as high as 80–95 percent, so maternal blood testing is recommended early in pregnancy.

Treatment of syphilis involves the administration of penicillin or other antibiotics.

Herpes virus: Genital herpes results from infection by herpes viruses. Two different viruses are involved. Eighty to ninety percent of genital herpes cases are caused by a specific virus known as **HV-2** (herpes virus Type 2), a virus usually associated with the genitalia. The remaining cases are caused by **HV-1**, the same virus responsible for cold sores on the mouth. Typically within a week of the initial infection the individual develops a number of painful, ulcerated lesions on the external genitalia. In women ulcers may also appear on the cervix. These gradually heal over the next 2–3 weeks. Recurring infections are common, although subsequent incidents are less severe.

Infection of the newborn infant during delivery with herpes viruses present in the vagina can lead to serious illness because the infant has few immunological defenses. Recent development of the antiviral agent *acyclovir* has helped treatment of initial infections.

 **PMS** *Page 940*

A number of physical and physiological changes occur in women 7–10 days before the start of menses. Fluid retention, breast enlargement, headaches, pelvic pain, and an uncomfortable "bloated" feeling are common complaints. These physical symptoms may be associated with psychological changes producing irritability, anxiety, and/or depression. This combination of symptoms has been called **premenstrual syndrome** (**PMS**). The mechanism responsible has yet to be determined. Changes in sex-hormone levels may be involved directly, by action on peripheral organ systems, or indirectly by modifying neurotransmitter release in the CNS. There are no laboratory tests or procedures to diagnose PMS, but following the appearance of symptoms over a 2–3 month period can reveal characteristic patterns. Treatment at present is symptomatic and may involve exercise, dietary change, and/or medication, depending on the nature of the primary symptom. For example, if headache is the major problem, analgesics are prescribed; diuretics may be used to combat bloating and fluid retention.

 **Endometriosis** *Page 942*

In **endometriosis** (en-dō-mē-trē-Ō-sis) an area of sloughed endometrial tissue reattaches and begins to grow outside the uterus. The severity of the condition depends on the size of the abnormal mass and its location. Abdominal pain, bleeding, pressure on adjacent structures, and infertility are common symptoms. As the island of endometrial tissue enlarges, the symptoms become more severe.

Diagnosis can usually be made by using a **laparoscope** (LAP-a-ro-skōp), a slender tube that is usually inserted through a small opening in the abdominal wall. Using this device a physician can see the outer surfaces of the uterus and uterine tubes,

144

the ovaries, and the lining of the pelvic cavity. Treatment may involve surgical removal of the endometrial masses or hormonal therapies. If the condition is widespread, a hysterectomy (removal of the uterus) and **oophorectomy** (removal of the ovaries) may be required.

CHAPTER 29
Development and Inheritance

Down Syndrome

Page 957

Individuals with **chromosomal abnormalities** have damaged, broken, missing, or extra copies of chromosomes. The normal human chromosomal complement, or *karyotype*, was described in Chapter 4 of the text and illustrated in Figure 4-15a. (p. 115) Chromosomal abnormalities produce a variety of serious clinical conditions, in addition to contributing to prenatal mortality. *Few fetuses with chromosomal abnormalities survive to full term.* The high mortality rate and severity of the problems reflect the fact that large numbers of genes have been added or deleted.

In general, individuals with extra or missing autosomal chromosomes do not survive. One notable exception involves the duplication of a particular autosomal chromosome, chromosome 21. Because there are three copies of this chromosome, rather than two, the condition is termed **trisomy**, specifically **trisomy 21**. Trisomy 21, also known as **Down syndrome**, is the most common chromosomal abnormality. Estimates of the frequency of appearance range from 1.5 to 1.9 per 1000 births for the U.S. population. The affected individual suffers from mental retardation and characteristic physical malformations, including a facial appearance that gave rise to the term *mongolism* once used to describe this condition. The degree of mental retardation ranges from moderate to severe, and few individuals with this condition lead independent lives. Anatomical problems affecting the cardiovascular system often prove fatal during childhood or early adulthood. Although some individuals survive to moderate old age, many develop Alzheimer's disease while still relatively young (before age 40). Typical features of an individual with Down syndrome are shown in Figure 29-A.

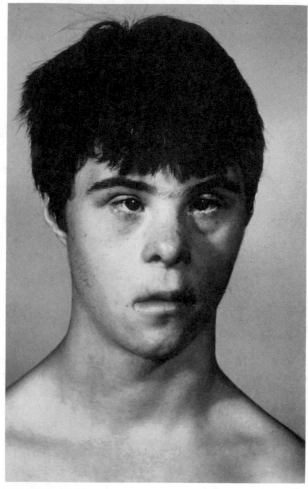

FIGURE 29-A
Down syndrome

For unknown reasons there is a direct correlation between maternal age and the risk of having a child with trisomy 21. Below maternal age 25 the incidence of Down syndrome approaches 1 in 2000 births, or 0.05 percent. For maternal ages 30–34, the odds increase to 1:900, and over the next decade they go from 1:300 to 1:100. For a mother over age 45 the risks are 1:50, or 2 percent. These statistics are becoming increasingly significant, for many women have delayed childbearing until their mid-30s or later.

Tests can be performed to determine whether or not a developing embryo suffers from trisomy 21 or from a number of less common chromosomal abnormalities. Positive results require careful deliberation on the part of the parents, as under these conditions current laws will permit induced abortions up to a gestational age of 4–5 months. (For a discussion of current testing methods, see Diagnostics: Chromosomal Analysis on p. 960 of the text.)

Sex Chromosome Abnormalities   *Page 959*

Abnormal numbers of sex chromosomes do not produce effects as severe as those induced by extra or missing autosomal chromosomes. In **Klinefelter syndrome** the individual carries the sex chromosome pattern *XXY*. The phenotype is male, but the extra X chromosome causes reduced androgen production. As a result the testes fail to mature, the individuals are sterile, and the breasts are slightly enlarged. The incidence of this condition among newborn males averages 1:750.

Turner's syndrome is caused by a sexual genotype abbreviated *XO*. This kind of chromosomal deletion is known as **monosomy**. The incidence of this condition at delivery has been estimated as 1:10,000. At birth the condition may not be recognized, for the phenotype is normal female. But maturational changes do not appear at puberty. The ovaries are nonfunctional, and estrogen production occurs at negligible levels.

Induction and Sexual Differentiation   *Page 962*

The physical (phenotypic) sex of a newborn infant depends upon the hormonal cues received during development, not on the genetic sex of the individual. If something disrupts the normal inductive processes, the individual's genetic sex and anatomical sex may be different. Such a person is called a **pseudohermaphrodite** (su-dō-her-MAF-ro-dīt). For example, if a female embryo becomes exposed to male hormones, it will develop the sexual characteristics of a male. Cases of this sort are relatively rare. The most common cause is the hypertrophy of the fetal adrenals and their production of androgens in high concentrations. Maternal exposure to androgens, in the form of anabolic steroids or as a result of an endocrine tumor, can also produce a female pseudohermaphrodite.

Male pseudohermaphrodites may result from an inability to produce adequate concentrations of androgens due to some enzymatic defect. In **testicular feminization syndrome** the infant appears to be a normal female at birth. Typical physical changes occur at puberty, and the individual has the overt physical and behavioral characteristics of an adult woman. Menstrual cycles do not appear, however, for the vagina ends in a blind pocket, and there is no uterus. Biopsies performed on the gonads reveal normal testicular structure, and the interstitial cells are busily

147

secreting testosterone. The problem apparently involves a defect in the cellular receptors sensitive to circulating androgens. Neither the embryo nor the adult tissues can respond to the testosterone produced by the gonads, so the person develops and remains physically a female.

If detected at infancy, many cases of pseudohermaphroditism can be treated with hormones and surgery to produce males or females of normal appearance. Depending on the arrangement of the internal organs and gonads, normal reproductive function may be more difficult to achieve.

Pseudohermaphroditism is an example of a developmental problem caused by hormonal miscues or an inability to respond appropriately to hormonal instructions. Another example is provided by male infertility associated with maternal exposure to a synthetic steroid, *diethylstilbestrol* (*DES*), prescribed in the 1950s to prevent miscarriages. At maturity an estimated 28 percent of male offspring produced abnormally small amounts of semen with marginal sperm counts. Daughters also have higher than normal infertility rates, due to uterine, vaginal, and uterine tube abnormalities, and they have an increased risk of developing vaginal cancer.

Problems with the Implantation Process   *Page 965*

The trophoblast undergoes repeated nuclear divisions, shows extensive and rapid growth, has a very high demand for energy, invades and spreads through adjacent tissues, and fails to activate the maternal immune system. In short, the trophoblast has many of the characteristics of cancer cells. In an estimated 0.1 percent of pregnancies, a definitive placenta does not develop, and instead the syncytial trophoblast runs wild, forming a **gestational neoplasm**, or **hydatidiform** (hī-da-TID-i-fōrm) **mole**. Prompt surgical removal of the mass is essential, sometimes followed by chemotherapy, for about 20 percent of hydatidiform moles will metastasize, invading other tissues with potentially fatal results.

Implantation usually occurs at the endometrial surface lining the uterine cavity. The precise location within the uterus varies, although most often implantation occurs in the body of the uterus. This is not an ironclad rule, and in an **ectopic pregnancy** implantation occurs somewhere other than within the uterus.

The incidence of ectopic pregnancies is approximately 0.6 percent. Women douching regularly have a 4.4 times higher risk of experiencing an ectopic pregnancy, presumably because the flushing action pushes the zygote away from the uterus. If the uterine tube has been scarred by a previous episode of pelvic inflammatory disease, there is also an increased risk of an ectopic pregnancy. Although implantation may occur within the peritoneal cavity, in the ovarian wall, or in the cervix, 95 percent of ectopic pregnancies involve implantation within a uterine tube. The

148

tube cannot expand enough to accommodate the developing embryo, and it usually ruptures during the first trimester. At this time the hemorrhaging that occurs in the peritoneal cavity may be severe enough to pose a threat to the woman's life.

In a few instances the ruptured uterine tube releases the embryo with an intact umbilical cord, and further development can occur. About 5 percent of these **abdominal pregnancies** actually complete full-term development; normal birth cannot occur, but the infant can be surgically removed from the abdominopelvic cavity. Because abdominal pregnancies are possible, it has been suggested that men as well as women could act as surrogate mothers if a zygote were surgically implanted in the peritoneal wall. It is not clear how the endocrine, cardiovascular, nervous, and other systems of a man would respond to the stresses of pregnancy. However, the procedure has been tried successfully in mice, and experiments continue.

Problems with Placentation   *Page 971*

In a **placenta previa** (PRĒ-vē-a; "in the way") implantation occurs in or near the cervix. This condition causes problems as the growing placenta approaches the internal cervical orifice. In a **total placenta previa** the placenta actually extends across the internal orifice, while a **partial placenta previa** only partially blocks the os. The placenta is characterized by a rich fetal blood supply, and the erosion of maternal blood vessels within the endometrium. Where the placenta passes across the internal orifice the delicate complex hangs like an unsupported water balloon. As the pregnancy advances, even minor mechanical stresses can be enough to tear the placental tissues, leading to massive fetal and maternal hemorrhaging.

Most cases are not diagnosed until the seventh month of pregnancy, as the placenta reaches its full size. At this time the dilation of the cervical canal and the weight of the uterine contents are pushing against the placenta where it bridges the internal orifice. Minor, painless hemorrhaging usually appears as the first sign of the condition. The diagnosis can usually be confirmed by ultrasound scanning. Treatment in cases of total placenta previa usually involves bed rest for the mother until the fetus reaches a size at which cesarean delivery can be performed with a reasonable chance of neonatal (newborn) survival.

In an **abruptio placentae** (ab-RUP-shē-ō pla-SEN-tē) part or all of the placenta tears away from the uterine wall sometime after the fifth month of gestation. The bleeding into the uterine cavity and the pain that follows usually will be noted and reported, although in some cases the shifting placenta may block the passage of blood through the cervical canal. In severe cases the hemorrhaging leads to maternal anemia, shock, and kidney failure. Although maternal mortality is low, the fetal mortality rate from this condition ranges from 30 to 100 percent, depending on the severity of the hemorrhaging.

Teratogens and Abnormal Development

Teratogens (TER-a-tō-jens) are stimuli that disrupt normal development by damaging cells, altering chromosome structure, or acting as abnormal inducers. **Teratology** (ter-a-TOL-o-jē) is literally the "study of monsters," and it considers extensive departures from the pathways of normal development. Teratogens that affect the embryo in the first trimester will potentially disrupt cleavage, gastrulation, or neurulation. The embryonic survival rate will be low, and the survivors will usually have severe anatomical and physiological defects affecting all of the major organ systems. Errors introduced into the developmental process during the second and third trimesters will be more likely to affect specific organs or organ systems, for the major organizational patterns are already established. Nevertheless, the alterations reduce the chances for long-term survival.

Many powerful teratogens are encountered in everyday life. The location and severity of the resulting defects vary depending on the nature of the stimulus and the time of exposure. Radiation is a powerful teratogen that can affect all living cells. Even the X-rays used in diagnostic procedures can break chromosomes and produce developmental errors; thus nonionizing procedures such as ultrasound are used to track embryonic and fetal development. Fetal exposure to the microorganisms responsible for syphilis or rubella ("German measles") can also produce serious developmental abnormalities, including congenital heart defects, mental retardation, and deafness.

Some chemical agents, especially those acting as abnormal inducers, will be teratogenic only if present at a time when embryonic or fetal targets have the competence to respond to them. There are literally thousands of critical inductions under way during the first trimester, initiating developmental sequences that will produce the major organs and organ systems of the body. In almost every case the nature of the inducing agent remains unknown, and the effects of unusual compounds within the maternal circulation cannot be predicted. As a result, virtually any unusual chemical that reaches an embryo has the potential for producing developmental abnormalities. Despite extensive testing with laboratory animals uncertainties remain, because the chemical nature of the inducer responsible for a specific process may vary from one species to another. For example, thalidomide produces abnormalities in humans and monkeys, but developing mice, rats, and rabbits are completely unaffected by the drug.

More powerful teratogenic agents will have an effect regardless of the time of exposure. Pesticides, herbicides, and heavy metals are common around agricultural and industrial environments, and these substances can contaminate the drinking water in the area. A number of prescription drugs, including certain antibiotics, tranquilizers, sedatives, steroid hormones, diuretics, anesthetics, and analgesics

150

also have teratogenic effects. During pregnancy the "Caution" label should always be consulted before using drugs without the advice of a physician.

Even practices usually considered socially acceptable for adults may not be acceptable to the fetus. Two examples are alcohol consumption and smoking. **Fetal alcohol syndrome (FAS)** occurs when maternal alcohol consumption produces developmental defects such as skeletal deformation, cardiovascular defects, and neurological disorders. Mortality rates in the newborn can be as high as 17 percent, and the survivors are plagued by problems in later development. The most severe cases involve mothers who consume the alcohol content of at least 7 ounces of hard liquor, 10 beers, or several bottles of wine each day. But because the effects produced are directly related to the degree of exposure, there is probably no level of alcohol consumption that can be considered completely safe. Fetal alcohol syndrome is the number one cause of mental retardation in the United States today, affecting roughly 7500 infants each year.

Smoking presents another major risk to the developing fetus. In addition to introducing potentially harmful chemicals, such as nicotine, smoking lowers the pO_2 of maternal blood and reduces the amount of oxygen arriving at the placenta. The fetus carried by a smoking mother will not grow as rapidly as one carried by a nonsmoker, and smoking increases the risks of spontaneous abortion, prematurity, and fetal death. There is also a higher rate of infant mortality after delivery, and postnatal development can be adversely affected.

Problems with the Maintenance of a Pregnancy

Page 975

The rate of maternal complications during pregnancy is relatively high. Pregnancy stresses maternal systems, and the stresses can overwhelm homeostatic mechanisms. The term **toxemia** (tok-SĒ-mē-a) **of pregnancy** refers to disorders affecting the maternal cardiovascular system. Chronic hypertension is the most characteristic symptom, but fluid balance problems and CNS disturbances, leading to coma or convulsions, may also occur. Some degree of toxemia occurs in 6–7 percent of third-trimester pregnancies. Severe cases account for 20 percent of maternal deaths and contribute to an estimated 25,000 neonatal (newborn) deaths each year.

Toxemia of pregnancy includes **preeclampsia** (prē-ē-KLAMP-sē-a) and **eclampsia** (ē-KLAMP-sē-a). Preeclampsia most often occurs during a woman's first pregnancy. Systolic and diastolic pressures become elevated, reaching levels as high as 180/110. Other symptoms include fluid retention and edema, along with CNS disturbances and alterations in kidney function. Roughly 4 percent of individuals with preeclampsia develop eclampsia. Eclampsia, or *pregnancy-induced hypertension* (*PIH*), is heralded by the onset of severe convulsions lasting 1–2 minutes followed by a variable period of coma. Other symptoms resemble those of preeclampsia, with additional evidence of liver and kidney damage. The mortality rate from eclampsia is approximately 5 percent; to save the mother the fetus must be delivered immediately. Once the fetus and placenta are removed from the uterus, symptoms of eclampsia disappear over a period of hours to days.

Common Problems with Labor and Delivery   *Page 978*

There are many potential problems during labor and delivery. Two relatively common types of complications are *forceps deliveries* and *breech births*.

By the end of gestation the fetus has usually rotated within the uterus so that it will enter the birth canal head first, with the face turned toward the sacrum. In around 6 percent of deliveries the fetus faces the pubis rather than the sacrum. Although these infants can eventually be delivered normally, risks to infant and mother increase the longer the fetus remains in the birth canal. Often the clinical response is the removal of the infant through a **forceps delivery**. The forceps used resemble a large, curved set of salad tongs that can be separated for insertion into the vaginal canal one side at a time. Once in place they are reunited and used to grasp the head of the infant. An intermittent pull is applied, so that the forces on the head resemble those encountered during normal delivery.

In 3–4 percent of deliveries, the legs or buttocks of the fetus enter the vaginal canal first. Such deliveries are known as **breech births**. Risks to the infant are relatively higher in breech births because the umbilical cord may become constricted, and placental circulation cut off. Because the head is normally the widest part of the fetus, the cervix may dilate enough to pass the legs and body but not the head. Entrapment of the fetal head compresses the umbilical cord, prolongs delivery, and subjects the fetus to severe distress and potential damage. If the fetus cannot be repositioned manually, a cesarian section is usually performed.

152

Cadaver
Dissection
Slides

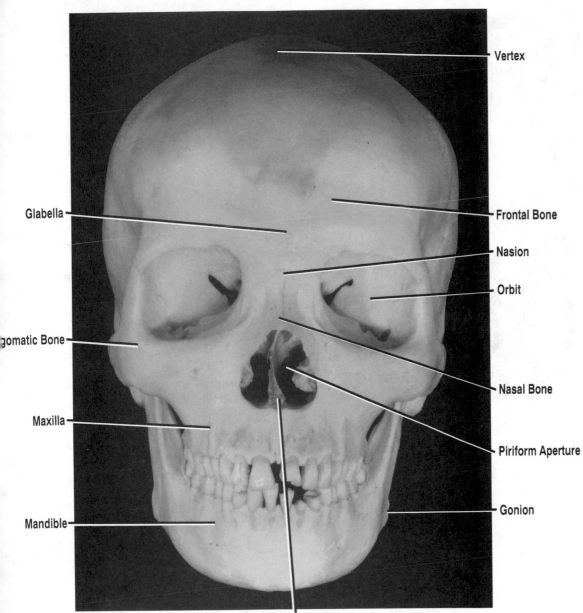

Vertex

Frontal Bone

Nasion

Orbit

Glabella

Nasal Bone

Zygomatic Bone

Maxilla

Piriform Aperture

Mandible

Gonion

Anterior Nasal Spine

155

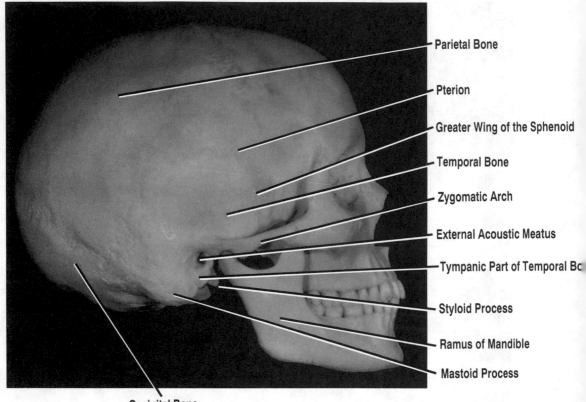

Parietal Bone

Pterion

Greater Wing of the Sphenoid

Temporal Bone

Zygomatic Arch

External Acoustic Meatus

Tympanic Part of Temporal Bc

Styloid Process

Ramus of Mandible

Mastoid Process

Occipital Bone

SLIDE 2
The Skull and Mandible: Right Lateral View

156

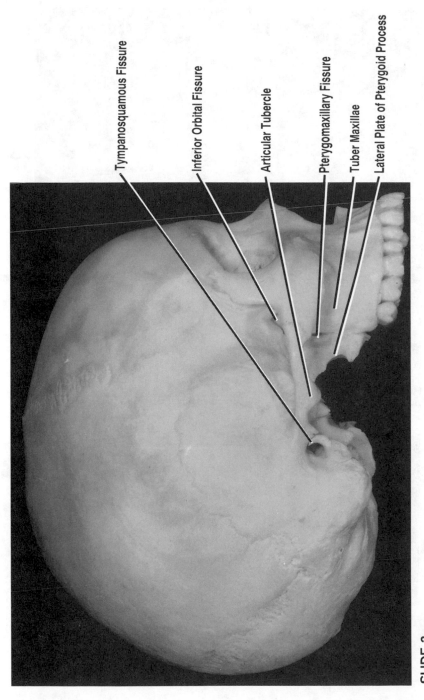

Tympanosquamous Fissure

Inferior Orbital Fissure

Articular Tubercle

Pterygomaxillary Fissure

Tuber Maxillae

Lateral Plate of Pterygoid Process

SLIDE 3
The Skull: Right Lateral View

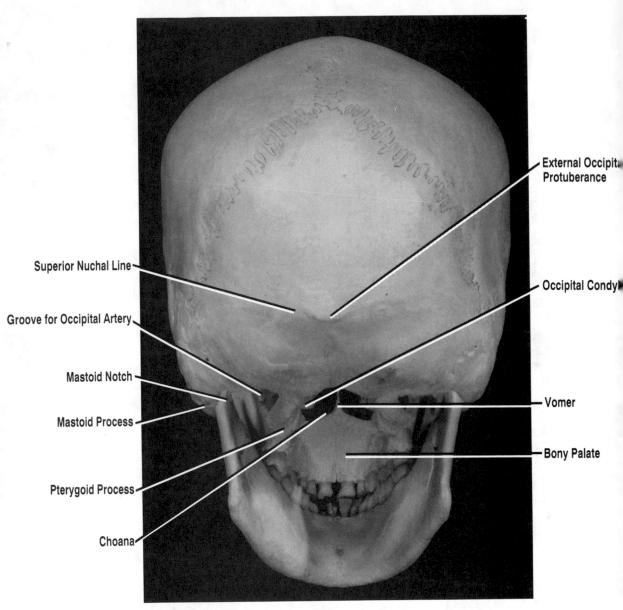

External Occipital Protuberance

Occipital Condyle

Superior Nuchal Line

Groove for Occipital Artery

Mastoid Notch

Mastoid Process

Vomer

Bony Palate

Pterygoid Process

Choana

SLIDE 4
The Skull and Mandible: Posterior View

158

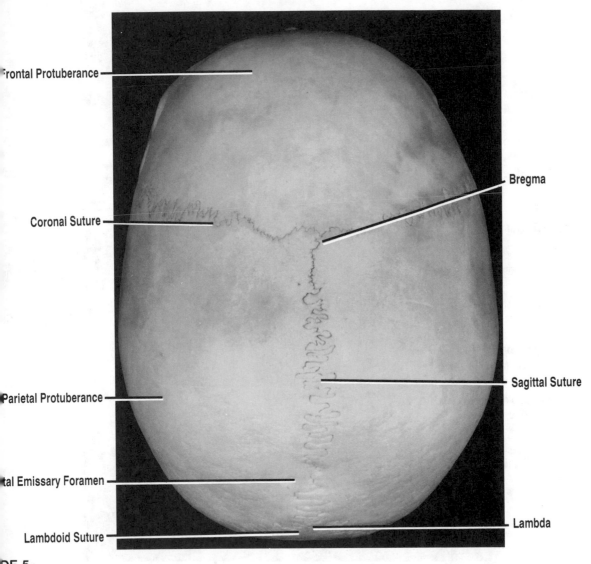

Frontal Protuberance

Coronal Suture

Bregma

Parietal Protuberance

Sagittal Suture

tal Emissary Foramen

Lambda

Lambdoid Suture

DE 5
e Skull: Posterior View

159

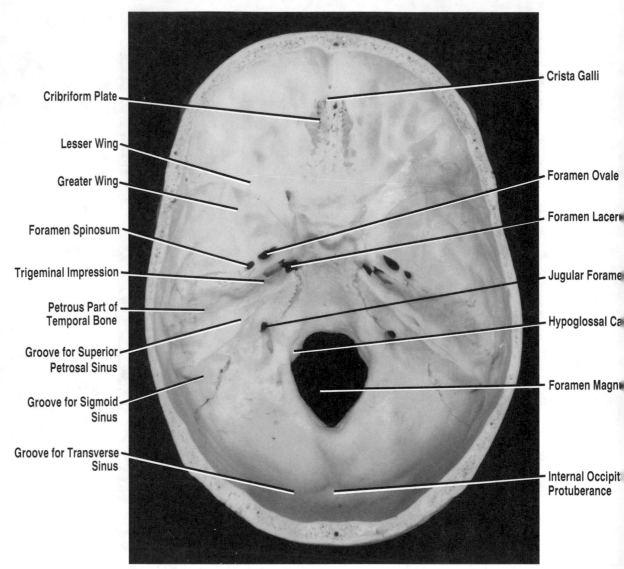

Cribriform Plate

Lesser Wing

Greater Wing

Foramen Spinosum

Trigeminal Impression

Petrous Part of
Temporal Bone

Groove for Superior
Petrosal Sinus

Groove for Sigmoid
Sinus

Groove for Transverse
Sinus

Crista Galli

Foramen Ovale

Foramen Lacer

Jugular Forame

Hypoglossal Ca

Foramen Magn

Internal Occipit
Protuberance

SLIDE 6
The Base of the Skull: Superior View

160

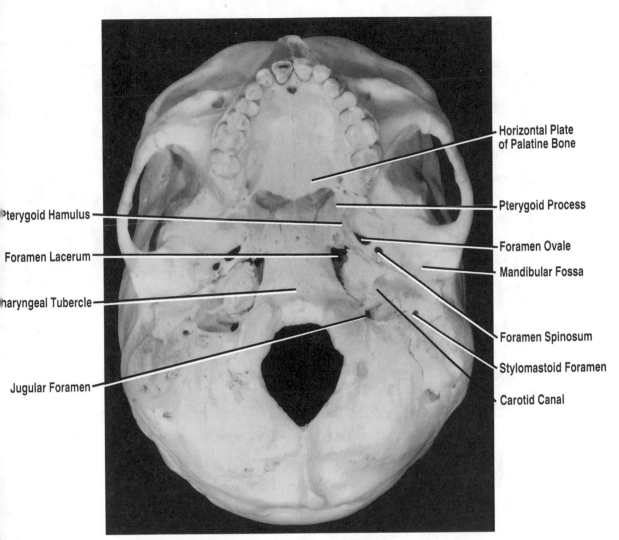

Horizontal Plate
of Palatine Bone

Pterygoid Process

Foramen Ovale

Mandibular Fossa

Foramen Spinosum

Stylomastoid Foramen

Carotid Canal

Pterygoid Hamulus

Foramen Lacerum

Pharyngeal Tubercle

Jugular Foramen

SLIDE 7
The Base of the Skull: Inferior View

161

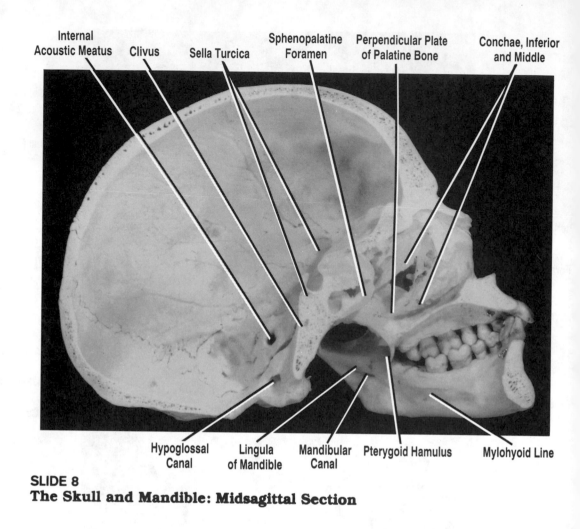

Internal Acoustic Meatus   Clivus   Sella Turcica   Sphenopalatine Foramen   Perpendicular Plate of Palatine Bone   Conchae, Inferior and Middle

Hypoglossal Canal   Lingula of Mandible   Mandibular Canal   Pterygoid Hamulus   Mylohyoid Line

SLIDE 8
The Skull and Mandible: Midsagittal Section

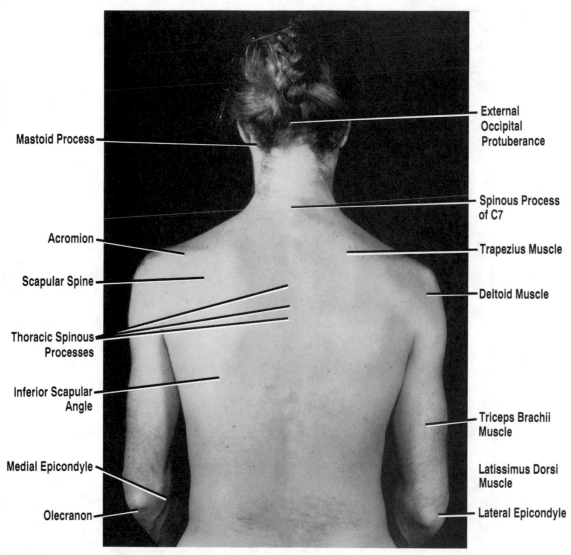

Mastoid Process

External Occipital Protuberance

Acromion

Spinous Process of C7

Scapular Spine

Trapezius Muscle

Deltoid Muscle

Thoracic Spinous Processes

Inferior Scapular Angle

Triceps Brachii Muscle

Medial Epicondyle

Latissimus Dorsi Muscle

Olecranon

Lateral Epicondyle

SLIDE 22
The Head, Neck, Thorax, and Arm of a Living Subject: Posterior View

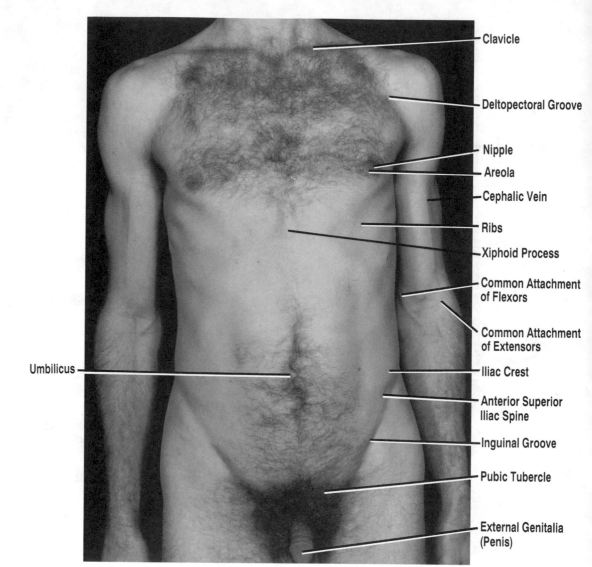

Clavicle

Deltopectoral Groove

Nipple

Areola

Cephalic Vein

Ribs

Xiphoid Process

Common Attachment
of Flexors

Common Attachment
of Extensors

Iliac Crest

Anterior Superior
Iliac Spine

Inguinal Groove

Pubic Tubercle

External Genitalia
(Penis)

Umbilicus

SLIDE 33
The Arm, Thorax, and Abdomen of a Living Male Subject: Anterior View

164

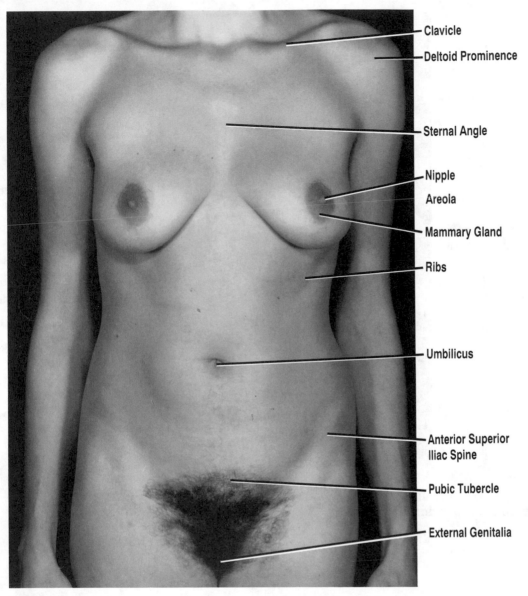

Clavicle

Deltoid Prominence

Sternal Angle

Nipple

Areola

Mammary Gland

Ribs

Umbilicus

Anterior Superior
Iliac Spine

Pubic Tubercle

External Genitalia

SLIDE 34
The Arm, Thorax, and Abdomen of a Living Female Subject: Anterior
View

165

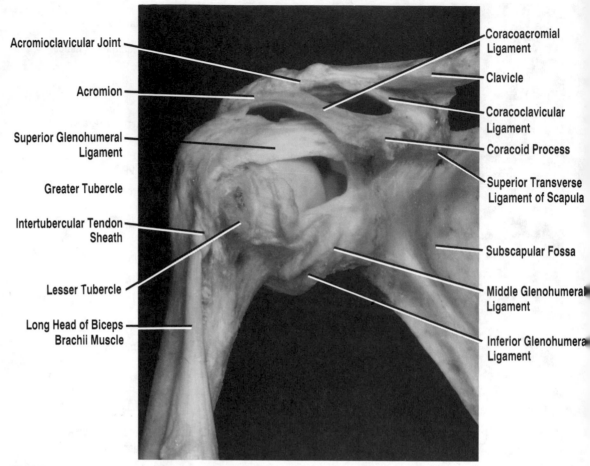

Acromioclavicular Joint

Acromion

Superior Glenohumeral Ligament

Greater Tubercle

Intertubercular Tendon Sheath

Lesser Tubercle

Long Head of Biceps Brachii Muscle

Coracoacromial Ligament

Clavicle

Coracoclavicular Ligament

Coracoid Process

Superior Transverse Ligament of Scapula

Subscapular Fossa

Middle Glenohumeral Ligament

Inferior Glenohumeral Ligament

SLIDE 39
The Right Glenohumural and Acromioclavular Joints and Associated Structures: Anterior View

166

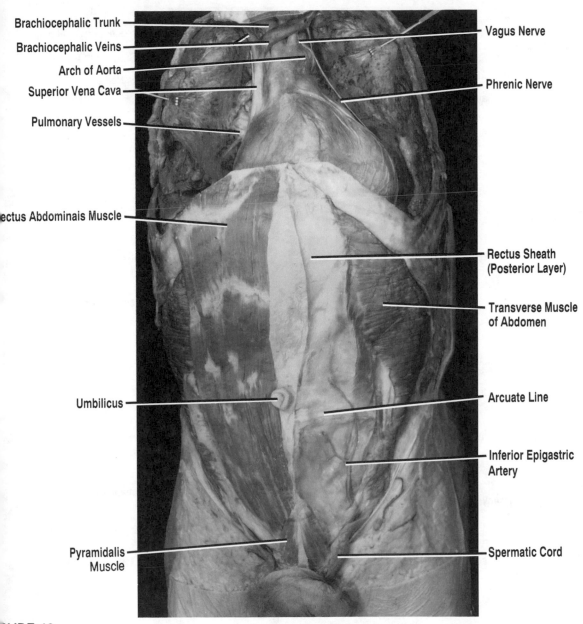

Brachiocephalic Trunk

Brachiocephalic Veins

Arch of Aorta

Superior Vena Cava

Pulmonary Vessels

Rectus Abdominais Muscle

Umbilicus

Pyramidalis
Muscle

Vagus Nerve

Phrenic Nerve

Rectus Sheath
(Posterior Layer)

Transverse Muscle
of Abdomen

Arcuate Line

Inferior Epigastric
Artery

Spermatic Cord

SLIDE 40
The Mediastinal Contents in situ and the Transverus Muscle of the
Abdomen

167

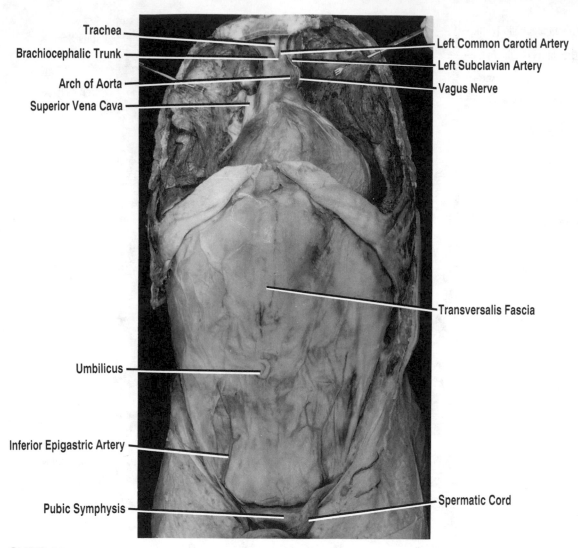

Trachea

Brachiocephalic Trunk

Arch of Aorta

Superior Vena Cava

Left Common Carotid Artery

Left Subclavian Artery

Vagus Nerve

Transversalis Fascia

Umbilicus

Inferior Epigastric Artery

Pubic Symphysis

Spermatic Cord

SLIDE 41
**The Arch of the Aorta and Branches in situ and the Transveralis Fascia:
Anterior View**

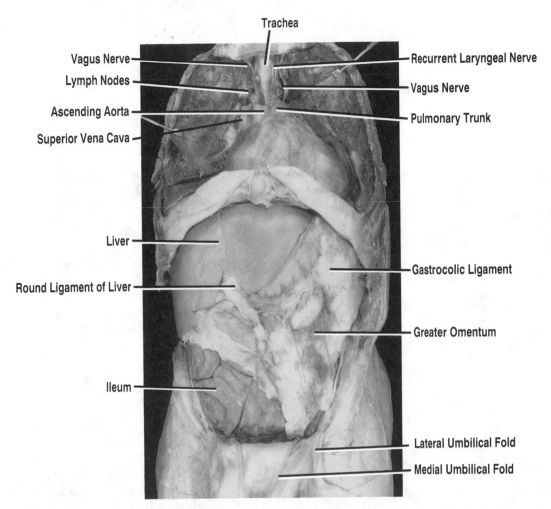

Trachea

Vagus Nerve

Lymph Nodes

Ascending Aorta

Superior Vena Cava

Recurrent Laryngeal Nerve

Vagus Nerve

Pulmonary Trunk

Liver

Round Ligament of Liver

Ileum

Gastrocolic Ligament

Greater Omentum

Lateral Umbilical Fold

Medial Umbilical Fold

SLIDE 42
The Mediastinal Airways and the Abdominal Organs in situ: Anterior View

169

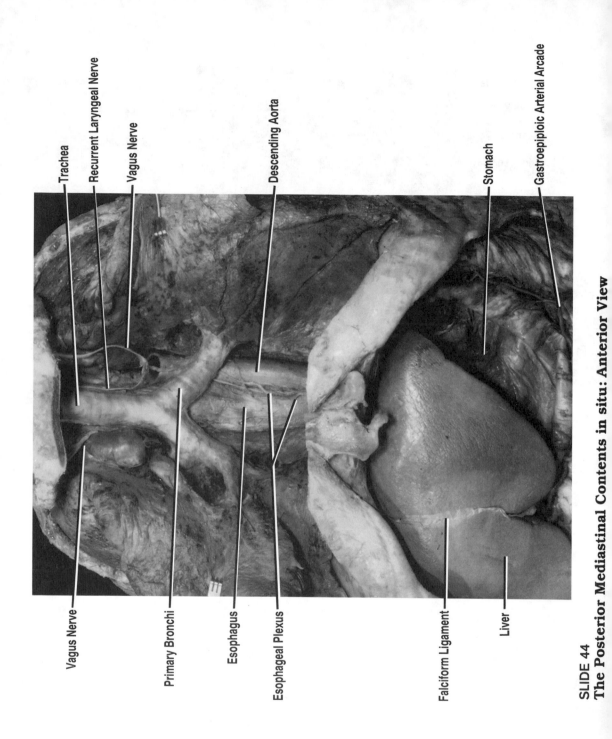

Trachea

Recurrent Laryngeal Nerve

Vagus Nerve

Descending Aorta

Stomach

Gastroepiploic Arterial Arcade

Vagus Nerve

Primary Bronchi

Esophagus

Esophageal Plexus

Falciform Ligament

Liver

SLIDE 44
The Posterior Mediastinal Contents in situ: Anterior View

170

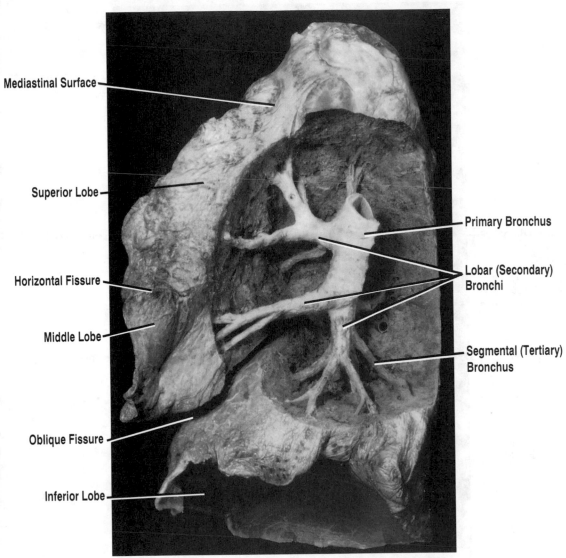

Mediastinal Surface

Superior Lobe

Horizontal Fissure

Middle Lobe

Oblique Fissure

Inferior Lobe

Primary Bronchus

Lobar (Secondary) Bronchi

Segmental (Tertiary) Bronchus

SLIDE 47
Bronchial Distribution on the Right Lung: Mediastinal View

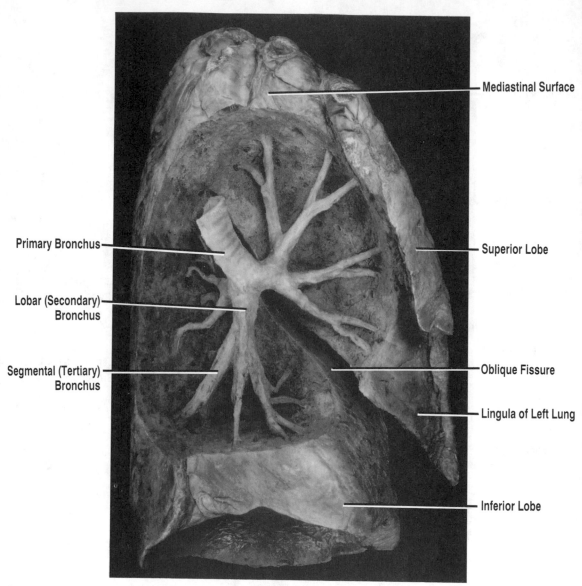

Mediastinal Surface

Primary Bronchus

Lobar (Secondary)
Bronchus

Segmental (Tertiary)
Bronchus

Superior Lobe

Oblique Fissure

Lingula of Left Lung

Inferior Lobe

SLIDE 48
Broncial Distribution of the Left Lung: Mediastinal View

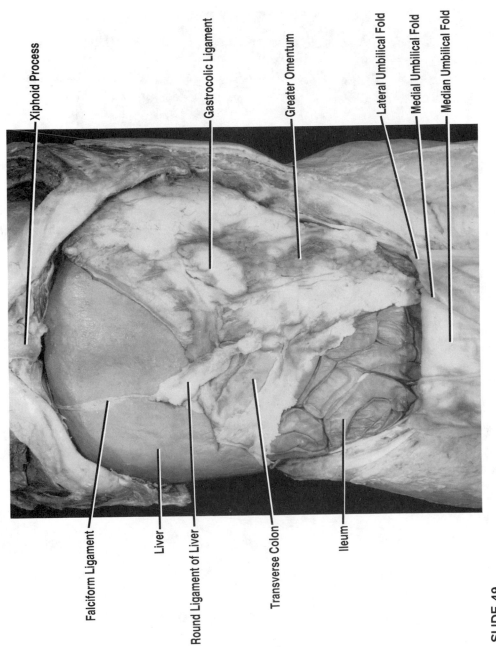

Xiphoid Process

Gastrocolic Ligament

Greater Omentum

Lateral Umbilical Fold

Medial Umbilical Fold

Median Umbilical Fold

Falciform Ligament

Liver

Round Ligament of Liver

Transverse Colon

Ileum

SLIDE 49
Abdominal Contents in situ: Anterior View

173

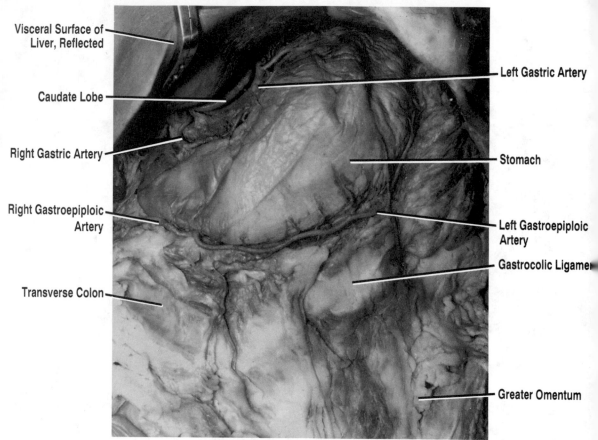

Visceral Surface of Liver, Reflected

Caudate Lobe

Right Gastric Artery

Right Gastroepiploic Artery

Transverse Colon

Left Gastric Artery

Stomach

Left Gastroepiploic Artery

Gastrocolic Ligament

Greater Omentum

SLIDE 50

The Stomach and Associated Blood Vessels in situ: Anterior View

174

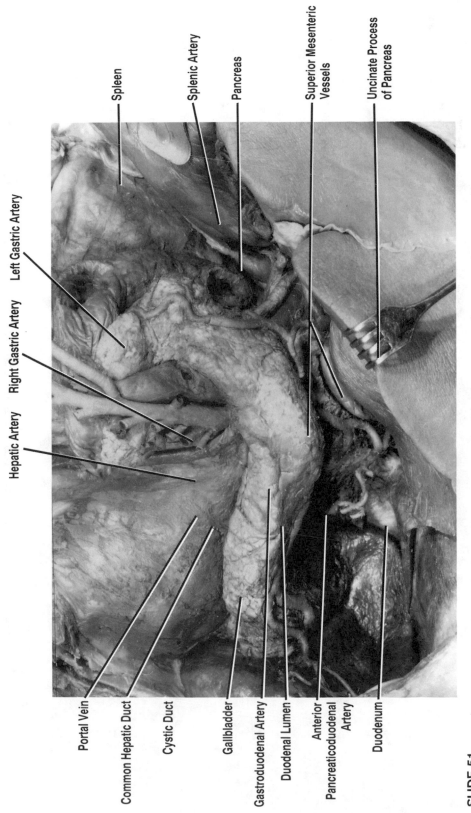

Spleen

Splenic Artery

Pancreas

Superior Mesenteric
Vessels

Uncinate Process
of Pancreas

Left Gastric Artery

Right Gastric Artery

Hepatic Artery

Portal Vein

Common Hepatic Duct

Cystic Duct

Gallbladder

Gastroduodenal Artery

Duodenal Lumen

Anterior
Pancreaticoduodenal
Artery

Duodenum

SLIDE 51
**The Pancreas, Spleen, Duodenum, and Associated Structures: Anterior
View**

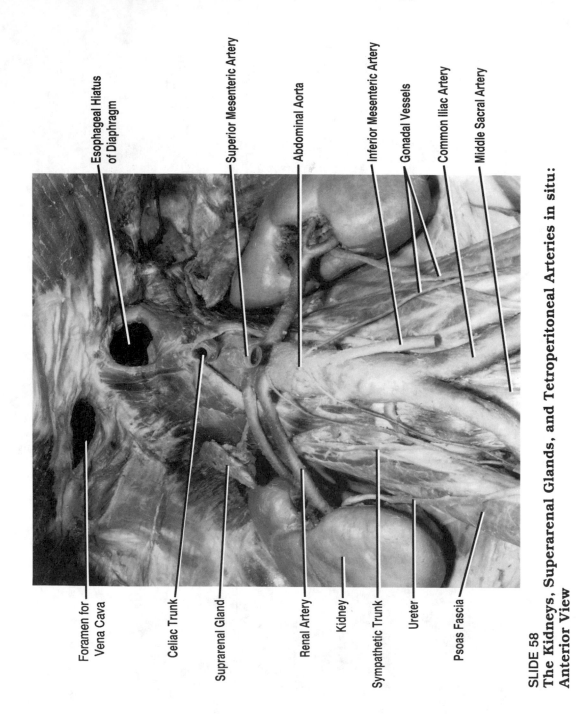

Esophageal Hiatus
of Diaphragm

Superior Mesenteric Artery

Abdominal Aorta

Inferior Mesenteric Artery

Gonadal Vessels

Common Iliac Artery

Middle Sacral Artery

Foramen for
Vena Cava

Celiac Trunk

Suprarenal Gland

Renal Artery

Kidney

Sympathetic Trunk

Ureter

Psoas Fascia

SLIDE 58
**The Kidneys, Superarenal Glands, and Tetroperitoneal Arteries in situ:
Anterior View**

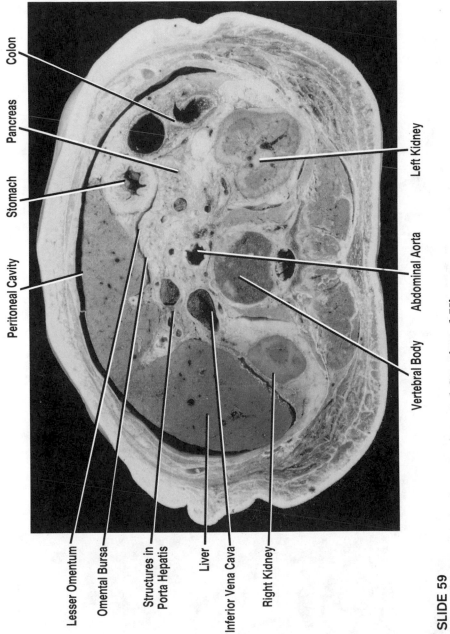

Colon

Pancreas

Stomach

Peritoneal Cavity

Left Kidney

Abdominal Aorta

Vertebral Body

Lesser Omentum

Omental Bursa

Structures in
Porta Hepatis

Liver

Inferior Vena Cava

Right Kidney

SLIDE 59
Abdominal Contents: Horizontal Sectional View

177

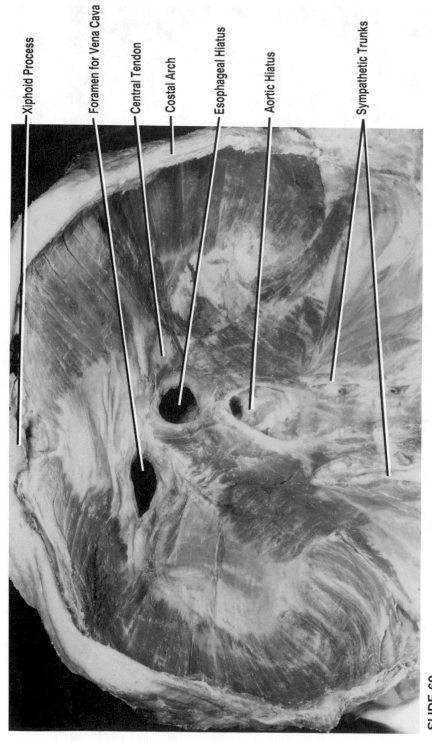

Xiphoid Process

Foramen for Vena Cava

Central Tendon

Costal Arch

Esophageal Hiatus

Aortic Hiatus

Sympathetic Trunks

SLIDE 60
The Diaphragm: Inferior View

178

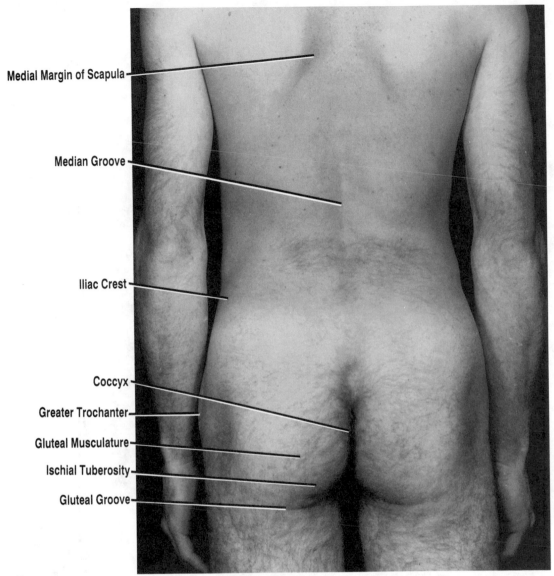

Medial Margin of Scapula

Median Groove

Iliac Crest

Coccyx

Greater Trochanter

Gluteal Musculature

Ischial Tuberosity

Gluteal Groove

SLIDE 61
The Lower Back and Gluteal Regions in a Living Subject: Posterior View

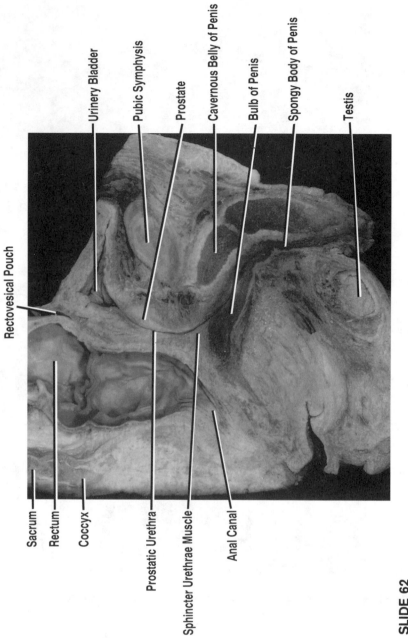

Rectovesical Pouch

Urinery Bladder

Pubic Symphysis

Prostate

Cavernous Belly of Penis

Bulb of Penis

Spongy Body of Penis

Testis

Sacrum

Rectum

Coccyx

Prostatic Urethra

Sphincter Urethrae Muscle

Anal Canal

SLIDE 62
The Male Pelvis, Perineum, and External Genitalia: Medial View of the Sagittal Section

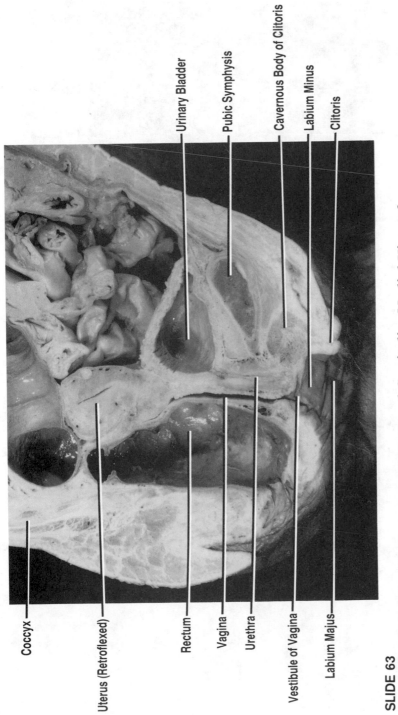

Coccyx

Uterus (Retroflexed)

Rectum

Vagina

Urethra

Vestibule of Vagina

Labium Majus

Urinary Bladder

Pubic Symphysis

Cavernous Body of Clitoris

Labium Minus

Clitoris

SLIDE 63
**The Female Pelvis, Perineum, and External Genitalia: Medial View of a
Sagittal Section**

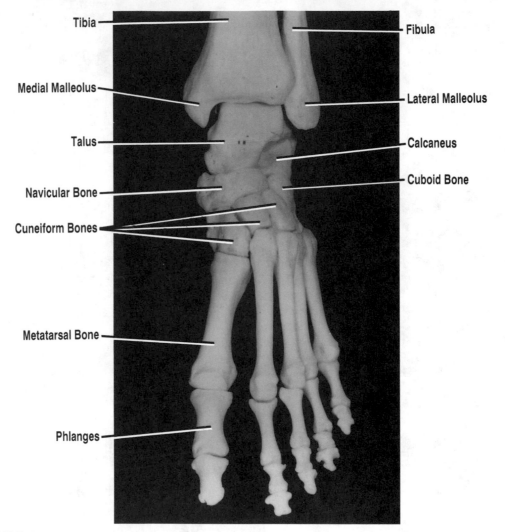

Tibia

Medial Malleolus

Talus

Navicular Bone

Cuneiform Bones

Metatarsal Bone

Phlanges

Fibula

Lateral Malleolus

Calcaneus

Cuboid Bone

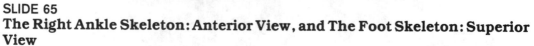

SLIDE 65
The Right Ankle Skeleton: Anterior View, and The Foot Skeleton: Superior View

182

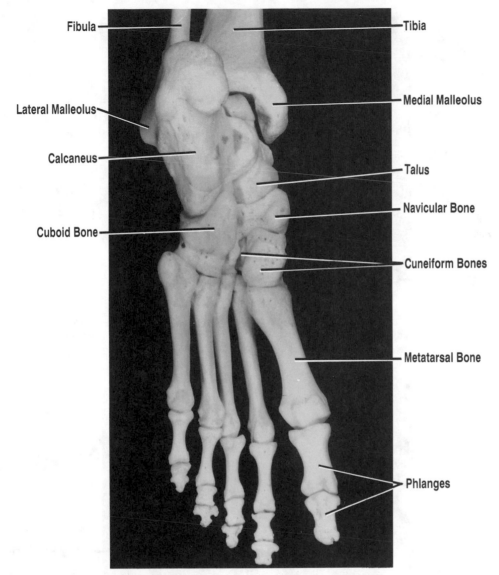

Fibula

Tibia

Lateral Malleolus

Medial Malleolus

Calcaneus

Talus

Navicular Bone

Cuboid Bone

Cuneiform Bones

Metatarsal Bone

Phlanges

SLIDE 66
The Right Ankle Skeleton: Posterior View, and The Foot Skeleton: Inferior View

183

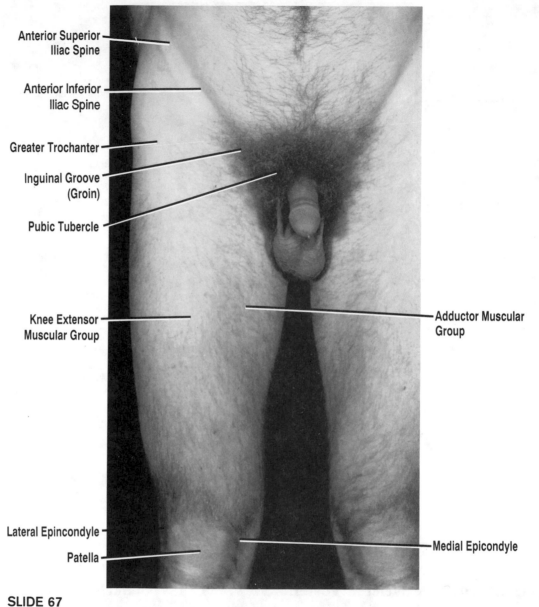

Anterior Superior Iliac Spine

Anterior Inferior Iliac Spine

Greater Trochanter

Inguinal Groove (Groin)

Pubic Tubercle

Knee Extensor Muscular Group

Lateral Epincondyle

Patella

Adductor Muscular Group

Medial Epicondyle

SLIDE 67
The Right Thigh of a Living Subject: Anterior View

184

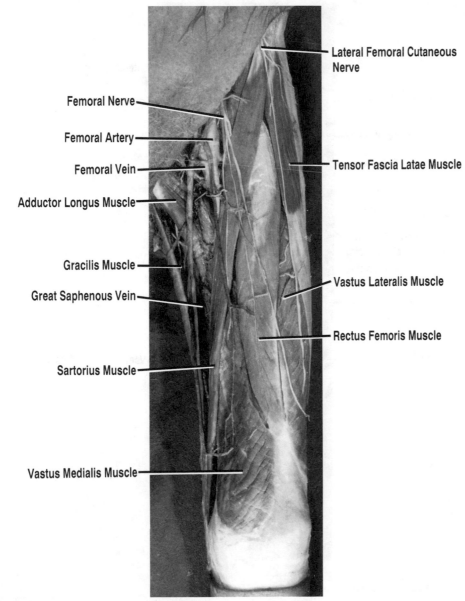

Femoral Nerve

Femoral Artery

Femoral Vein

Adductor Longus Muscle

Gracilis Muscle

Great Saphenous Vein

Sartorius Muscle

Vastus Medialis Muscle

Lateral Femoral Cutaneous Nerve

Tensor Fascia Latae Muscle

Vastus Lateralis Muscle

Rectus Femoris Muscle

SLIDE 68
Superficial Muscles of the Anterior and Medial Compartments of the Right Thigh

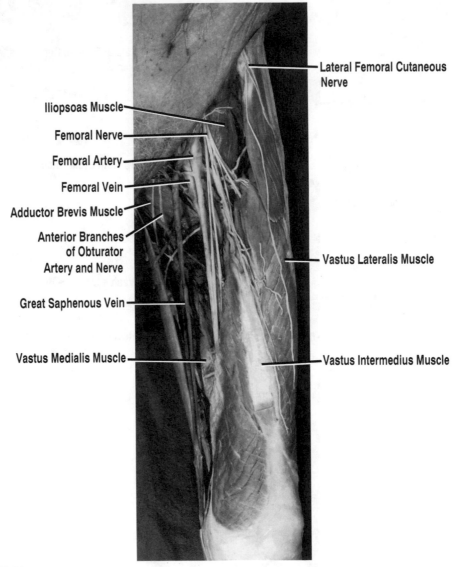

Lateral Femoral Cutaneous Nerve

Iliopsoas Muscle

Femoral Nerve

Femoral Artery

Femoral Vein

Adductor Brevis Muscle

Anterior Branches of Obturator Artery and Nerve

Great Saphenous Vein

Vastus Lateralis Muscle

Vastus Medialis Muscle

Vastus Intermedius Muscle

SLIDE 69
Femoral Triangle and Adductor Canal

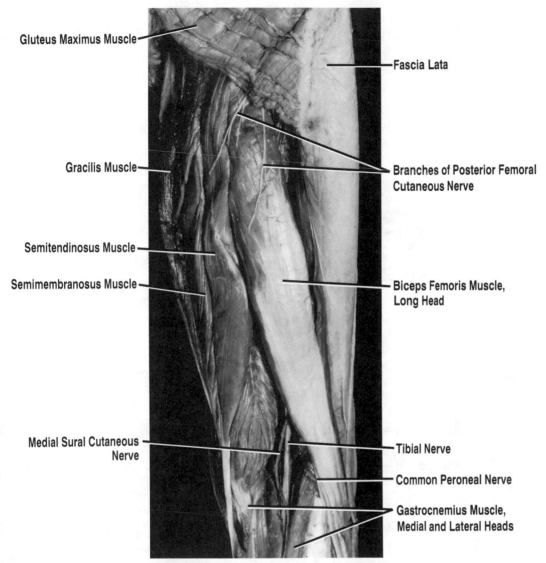

Gluteus Maximus Muscle

Gracilis Muscle

Semitendinosus Muscle

Semimembranosus Muscle

Medial Sural Cutaneous
Nerve

Fascia Lata

Branches of Posterior Femoral
Cutaneous Nerve

Biceps Femoris Muscle,
Long Head

Tibial Nerve

Common Peroneal Nerve

Gastrocnemius Muscle,
Medial and Lateral Heads

SLIDE 70
Superficial Structures in the Posterior Compartment of the Right Thigh

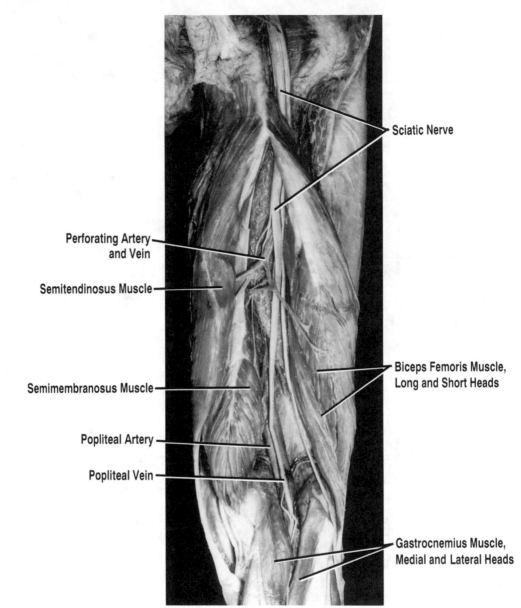

Sciatic Nerve

Perforating Artery
and Vein

Semitendinosus Muscle

Semimembranosus Muscle

Popliteal Artery

Popliteal Vein

Biceps Femoris Muscle,
Long and Short Heads

Gastrocnemius Muscle,
Medial and Lateral Heads

SLIDE 71
Deep Structures in the Posterior Compartment of the Right Thigh

SURVIVAL AND BEYOND

Success Strategies
for the Student
of Anatomy
and Physiology

Study Hints

Studying any subject, especially one as complex as human anatomy and physiology, is not a hit-or-miss proposition. There are systematic habits and skills that can be developed to greatly increase the efficiency of the studying process. Some of these are listed below.

• Make sure that you choose a suitable place for study. A comfortable desk or chair with good, over-the-shoulder lighting is ideal. A flat surface is useful to have—a desk or table top of comfortable height will do nicely. There should not be a lot of distractions in your study area. Loud talking, televisions, and blaring radios distract from the learning process.

• Establish a regular schedule of study. You should set aside thirty to sixty minutes each day for study. This is far better than trying to study for three hours at a stretch on one or two days of the week. As you will learn during your studies, neural synapses fatigue when used continuously for extended periods of time. Once that happens you are accomplishing nothing in terms of learning.

• Start each study session by quickly reviewing the previously covered material. The key to mastery is repetition. A quick review within 24 hours of having studied new material will greatly reinforce what was learned. A very easy way to do this is to read through the study outline that is provided at the end of each chapter of your textbook.

• Develop a systematic approach to studying. Science is a systematic way of looking at the world, and your approach to learning science should also be systematic. The approach detailed below is one that has proven very useful to a large number of students. References to specific features of *Fundamentals of Anatomy and Physiology* have been included to help you get the most out of your text.

1. If you are studying this material as part of a formal college course then you will in all probability be attending formal lectures. By all means be sure that you have read the chapter before attending lecture. You will find that the lecture is much more meaningful to you if you have read the material that is to be discussed.

2. Begin each chapter by first reading the chapter objectives at the start. These will tell you what the author considers to be the most important points.

3. Read the chapter in the textbook.

> • As you read the chapter pay particular attention to the terminology. One of the things that makes anatomy and physiology difficult for the new student is the extensive vocabulary needed to describe all its aspects. You will be introduced to more new terms in this course than are found in the typical foreign language course. They must be mastered, and you should begin immediately.

> • Also make sure that you pay attention to the illustrations and tables. A large amount of information is contained in these, and to bypass them eliminates about fifty percent of the information in the textbook.

> • When you come to the Concept Checkpoints that occupy key locations in each chapter, take a few minutes to answer the questions. They are designed to serve as intellectual "speed bumps"— they will tell you whether you have been reading too fast or not paying enough attention to what you have been reading. Some test chiefly memory, others require a bit of thought, but if you have been concentrating on what you have read you should be able to answer all of them quickly and easily. If you find that you are stumped, you probably need to reread the preceding material. (You can check your answers in Appendix VII of the text.)

> • Where recall of material covered in previous chapters is important for understanding new material, cross-references have been provided. Blue "chain-links" icons are used to call attention to these references, and to remind you that very few things about the human body can be studied in isolation—every topic is linked

to other topics. Whenever you see the links icon, stop for a moment to make sure you remember the material that is referred to. If not, go back and refresh your memory.

• Topics in this Clinical Manual are referenced in the text at the point where the material is most relevant. The title of the discussion follows the red caduceus icon and the letters CM. (The Clinical Manual topics for each chapter are also listed in the Review Planner for that chapter, and those for the entire text appear on the rear endpapers, along with a list of Boxes in the text.) Your instructor may assign some of these; you may wish to read others to shed additional light on the text material, or because you are interested in the topic.

4. After you have completed reading the text, be sure to read the study outline at the chapter's end. This study outline is a detailed summary of all of the major topics and vocabulary included in the chapter. It constitutes an excellent review, and is one of the outstanding features of your textbook.

5. Answer the end-of-chapter questions, using the Review Planner to organize your work. Each chapter objective is repeated in the Review Planner, along with a list of the Level 1 questions that are most relevant to that particular objective. Thus you can test your mastery of the chapter objectives one by one and identify any that you are still having difficulty with. The Level 2 questions call for more extended answers; they require an understanding of concepts and the ability to synthesize ideas. Level 3 questions help you to develop your critical thinking skills and allow you to apply your knowledge to actual clinical situations.

The Study Guide for your text contains many additional questions and exercises to help you master the material of this course.

Finally, do not become discouraged. To the beginning student the multitude of mysterious polysyllabic terms that characterizes human anatomy and physiology can be intimidating. Keep in mind that with a systematic approach to study and sufficient time, any average person can master the subject. The secret is to stay with it. As you proceed through the subject you will find that mastery of previous material makes the new material much easier, and eventually you will develop a "critical mass" of information that will permit you to move forward with confidence. The rewards of success are worth the time and energy. Besides all of the practical applications, there is a great deal of satisfaction in understanding how your body is constructed and how it functions.

The sections that follow, prepared by Professor Jeff Smith of Delgado Community College in New Orleans, provide more detailed discussions of techniques you may find useful for reading your text, taking notes in lecture, and studying.

Reading a Textbook

by Jeffrey L. Smith
Delgado Community College

How do you read a textbook? Simple. You sit at a desk or in a comfortable chair and read it, right? Probably not. Do you do any of the following things when reading a textbook?

Hope for many large pictures and graphs because they take up space that could be used for words?

Continually flip several pages ahead to see how close you are to the end of the chapter?

Stare at a page for several minutes before discovering that your mind was somewhere else?

Look back at the previous page and discover that you read it but don't remember any of it?

These are pretty good indicators of someone whose mind is drifting elsewhere. Authors write novels expecting you to read them in comfortable chairs. They're filled with drama, romance, suspense, and intrigue that keeps you turning pages to find out what happens next. If a novel doesn't hold your attention, you simply stop reading it. Textbooks have facts and concepts. Because a course textbook is rarely an exciting page-turner, you must develop techniques to keep your attention centered on your task.

Most students prefer not to read a chapter before the professor covers the material in class. To be honest, it _is_ probably easier to read the chapter after having heard about it first in lecture. However, it is much more _efficient_ to listen to the lecture having already covered the material in the book. If you get bogged down in the reading, you can always go back, slow down, or take a break. If you get lost in lecture, there is no going back. _Always read the chapter before the professor lectures about it._ You are eventually going to read it several times; why not read it once a few days earlier? It will make your lecture material much easier to understand.

There are several formulas for learning from textbooks. The following pages describe two such techniques: **SQ3R** and **SOAR**.

The SQ3R Study Formula

The five-step **SQ3R** formula (Survey, Question, Read, Recite, Review) may, at first, appear too time-consuming. _Try_ it, keeping in mind that you'll be using your study time much more efficiently. You won't have to reread the chapter as many times, and you will save time in the future.

Below are the five steps to **SQ3R**:

Survey

This consists of _previewing_ the material before you begin to study it. In the first few minutes before beginning to read the chapter, do the following steps:

Read the title of the chapter.

Read the chapter outline (usually at the end of the chapter) if one is present.

Read the introduction or first few paragraphs at the beginning of the chapter. This gives you a general overview of what the chapter is about.

Thumb through the entire chapter, page by page.

• Read the headings of sections. Some of these should be familiar. Don't worry if some headings seem foreign to you.

• Read the first sentence of each paragraph.

•Glance over the pictures, tables, and diagrams. Again, if they don't make sense now, don't worry about it.

•Read the summary or concluding paragraph at the end of the chapter.

That's it. In about ten minutes, you have become much more familiar with the chapter. Like seeing a preview of a movie or a television show, you now know something about the chapter. You have some idea of how long the chapter is, how easy or difficult it might be, and how you might want to break it up into smaller pieces. You can *double* your comprehension later by first getting a vague familiarity with the information to be studied.

Question

It is much easier to study material if you are doing it with the intention of answering questions. If you have an idea of what questions you are trying to answer, the material in the chapter will have more meaning. If they are available, read over the *Learning Objectives* for that chapter. If there are study questions at the back of the chapter, look at these as well. If you have some idea of what to be looking for, the important points will be more likely to stand out when you reach them during your reading.

Read

It is easier to master material that is in smaller pieces. Twenty years ago, many textbooks consisted of twenty chapters of over eighty pages each. Today, most textbooks have *more* chapters with *fewer* pages each. If the material you have reviewed in steps one and two (Survey and Question) appeared difficult, break up the chapter further into even smaller pieces *before* you start reading. For example, if it is a difficult or unfamiliar 20-page chapter, try to break it into four or five smaller subsections of 4-5 pages apiece.

Read the chapter or section with the idea of finding the answers to those questions and objectives you saw earlier. You will get much more out of your reading now that you have an idea of what you are looking for.

Don't simply *read* the chapter or section. You *must* use strategies for keeping your attention focused on the material. This is particularly important if you aren't feeling alert. Specifically, try one or more of the following:

Outline the material. Any time you *write* about something (versus simply reading it), you are much more likely to retain it. This takes longer than underlining, but is much more effective. To outline, you *must* actively think about what you are reading.

Make flash cards. This doesn't take as much time as outlining, and it forces you to think about potential questions.

Write answers to the learning objectives or study questions. Try to use your own wording. Why not just run through the answers in your head? (1) You may not realize that you can't answer the question until you try, and (2) you are much more likely to remember something that you have written.

Should you underline or highlight important sentences or phases? This requires the least effort, and is probably better than just reading alone. However, students who highlight their books often make the following fundamental errors:

Highlighting unimportant material. Many students have little idea of <u>what</u> to highlight.

Highlighting familiar material while skipping unfamiliar material. There is a tendency to assume something that you already know must be more important. However, focusing on the familiar material means you aren't learning anything new.

Highlighting with the idea of "I will study this later." Reading the chapter again later leads to more highlighting until most of the pages are yellow, pink, and blue. This makes the chapter even harder to read, and has wasted much time.

Do NOT try to answer questions by simply thumbing through the book's index and searching for answers. At best, this will result in piecemeal memorization of words and phrases without understanding their meaning or significance. Taking this unfortunate shortcut is very common because it gives students the mistaken impression that they are making progress in learning the material.

Once you have completed a section, don't immediately go on to the next section. Give the material a few minutes to settle in. The idea here is to master smaller pieces of the textbook without skipping over challenging material.

Recite

Although this literally means audibly *saying* the answers to questions (quietly to yourself or to another person), you might get the same benefit by writing the answers down on paper. Go through the learning objectives and try to answer the questions. Review your flash cards, and try to recite answers to them aloud or on paper. Don't simply think of the answer...*say* it or *write* it. Studies have shown that retention of material is much greater if you say or write the answers to questions rather than simply reading them. If you can't say it or write it, you probably don't know it. Many students claim that they "go blank" on exams. It is likely that they would have "gone blank" during this step as well. If you have learning objectives, it would be a *very good* idea to answer them *in writing* during reciting. There is a Latin proverb that applies here: *qui scribit bis legit*, or *he who writes reads twice*. For centuries, scholars have recognized that *writing out something* is very effective when trying to learn a subject thoroughly.

Review

The brain is very proficient at *forgetting* one-time events. You are most likely to retain information in long term memory if you refresh your memory at regular intervals. Take several minutes for a break, and then review what you have just covered. Read what you underlined, review your outlines, or quiz yourself with the flash cards. You often get brief breaks between other activities (work, school, home, or on the bus). Use this time to review the study questions or learning objectives and the answers you have written for them.

Mastering any new learning technique takes time, practice and refinement, but is worth the effort in the future. Don't give up on the **SQ3R** formula without really trying it in earnest for a few weeks. On the first few tries, you will be thinking about both the technique *and* the material in the book. Once you get accustomed to the technique, you will be able to devote all of your attention to the reading.

The secret to successful learning is keeping your mind actively involved in the material while avoiding distractions. It *is* difficult to keep your mind from wandering when reading a textbook. **SQ3R** gives you a specific purpose for your studying. It may seem like too much trouble, but the time you invest in studying will be much more efficient. Thus, in the future you will have more time for other activities, and you will be less frustrated.

The SOAR Study Formula

Like **SQ3R,** the **SOAR** formula (<u>S</u>urvey, <u>O</u>rganize, <u>A</u>nticipate, <u>R</u>ecite, and <u>R</u>eview) is a logical, step-wise method for studying textbooks. Remember: reading is not studying. It is possible to sit and read something and not retain any of it. The **SQ3R** and **SOAR** formulas are techniques for *studying* a textbook, and should help anyone retain more of what they are studying. **SOAR** may be even better than **SQ3R** for students of Anatomy and Physiology. Below is a summary of the four steps of **SOAR**.

Survey

Like **SQ3R, SOAR** begins with a general survey of the chapter. The survey steps are the same:

Read the *title of the chapter.*

Read the *chapter outline* (if your text provides them).

Read the *introduction or first few paragraphs* at the beginning of the chapter.

Thumb through the entire chapter, page by page.

•Read the *headings of sections.*

•Read the *first sentence* of each paragraph.

•Glance over the *pictures, tables,* and *diagrams.*

•Read the *summary* or *concluding paragraph* at the end of the chapter.

Organize

Well-organized information is much easier to remember. There are several different ways to organize the information in the chapter as you read it. You should try each until you find a particular method that works best for you.

Read the chapter and *outline it* or make *concept maps* of the information. When possible, avoid making *long* lists; always try to limit lists to about five subjects by subdividing into categories.

Read the chapter and *take notes,* either in a spiral-bound notebook or on note cards.

Again, *underlining* and *highlighting* "important" material may seem more effective than it really is. Unless you are certain that this method works well for you, *don't* use highlighters.

Anticipate

The word *anticipate* here refers to successfully predicting what an instructor might ask on an exam. In effect, you place yourself in the role of a teacher. If your instructor provides learning objectives, this step is simple; he or she will ask questions that test your understanding of those objectives. Students who consistently make high grades on exams can often look at paragraph in a text and say *"That looks like something my instructor would ask."* Students who cannot imagine what an instructor would ask must resort to trying to learn everything. Soon, they get frustrated and quit. The student who can spot *major* ideas and differentiate them from *examples* (supporting information) has much less material to remember. The learning objectives should help you with the task of anticipating questions.

Merely anticipating the questions is half the task. You must prove to yourself that you can correctly answer the questions. Making flash cards on 3" x 5" or 5" x 8" index cards is an excellent technique for review or study of unfamiliar terminology. Do NOT simply use other people's cards; you *must* learn how to anticipate questions on your own. If you are having trouble, talk to your instructor.

Recite and Review

Instead of *Recite* and *Review*, many students make the mistake of trying *Read* and *Reread* as their study strategy. It is probably a *better* idea to go through the chapter with the idea that it will be the last time you look at the material. If you have correctly followed the previous steps, you should have some written materials such as notes, outlines, or flash cards that you can study. You cannot learn effectively by simply reading alone.

> Recite — to recite literally refers to saying what you have learned aloud. As silly as this sounds, if you recite what you have learned aloud, you are much more likely to remember it later. You can either say it to yourself, to a study partner, or anyone who is willing to listen. The important thing is that you can repeat the information, quietly, aloud, or on paper, without reading it.

> Review — the more you review something (that is, the more times you must think about it), the more likely it will stay in your memory. Your brain is particularly adept at forgetting even interesting information if you don't reinforce that information through repetition. To move information from short-term memory to long-term memory, you need to

refresh the circuits in your brain by reviewing previously-studied material at reasonable intervals. Look over what you have highlighted or underlined, read your chapter outlines, review your flash cards and set aside those that you have really mastered. Always give yourself enough

lead time to allow for review of material at a comfortable pace. Study ahead of time, *some every day*, and you should be able to spend the evening before an exam reciting and reviewing only.

An outstanding form of review is to place yourself in the role of teacher. The reason is well expressed by the Latin saying, *qui docet discit—he who teaches, learns.* If you have a study partner in the class, review the material on a regular basis, and place yourselves in the role of teacher rather than student. Make a commitment to know it well enough to explain it to someone who doesn't know it. If you can do this, you will have no difficulty on an exam. Too many students try only to learn it well enough to fake it on a multiple-choice test.

The two methods for reading a textbook described above should help you focus your attention on the material and should make the time you spend reading much more efficient. Try the method that seems to fit your learning style and personality the best, and feel free to alter some steps as you see fit.

Effective Note-Taking

Lecture notes are frequently your best source of information when studying for exams. When an instructor covers material in lecture, you should assume that it must be more important than material that he or she omitted. Nevertheless, effective note-taking is a neglected skill. Each semester, many dejected students go to their instructor's office for help. Although they often claim to have taken usually voluminous lecture notes, those notes often amount to pages of disjointed words and diagrams copied verbatim from the chalkboard. Many professors refuse to use the chalkboard at all; students who daydream in those classes often *have no lecture notes at all.* Taking poor notes may be a worse curse than having no textbook.

Why is effective note-taking such a neglected learning strategy? There are several reasons:

Poor listening skills — People tend to *listen* (not just *hear*, but really devote their attention) to someone speaking for periods of about thirty seconds, followed by variable periods of inattention. *Many* students become alert *only* when the professor writes something on the chalkboard. After they mindlessly copy the few words into their notebooks, their attention wanders off until the professor uses the chalkboard again.

Selective note-taking — There is a tendency to write down what you want to hear or expect to hear, *e.g.,* material that you already knew before taking the class. This may also pose a problem when these students underline or highlight their books; they underline or highlight the material they already know, and ignore unfamiliar material. It would make more sense to write down only the unfamiliar material.

Laziness — If writing complete lecture notes seems like too much effort, then maybe you aren't setting high enough goals for yourself. When you are actively taking good lecture notes, a fifty-minute class period seems like twenty minutes.When you are daydreaming and staring at the clock, it seems like *hours.*

How can you improve your note-taking skills? That depends on your current note-taking skills. If you are a slow writer, you should go straight to a department store and buy a portable tape recorder now. You may not need it later, but it is always better to have a taped lecture as insurance if you get bogged down or lost in lecture.

Even with the lecture taped, you need to develop effective note-taking skills. Keep the following in mind:

You <u>cannot</u> write every word that your instructor says. If you attempt to write down every word, you will succeed only in writing down the first five words of several hundred sentences.

You <u>cannot</u> listen to your instructor and paraphrase the lecture into complete sentences. To do so, you would have to hear the complete sentence, think about it, paraphrase it, and then begin to write it. While doing that, you are missing the next two sentences.

To take effective notes, you are going to have to come to class prepared. This means familiarizing yourself with the lecture topic <u>before</u> your professor covers it in lecture. For some inexplicable
reason, many students obstinately resist this suggestion. However, the same students invariably read the chapter a day or two *after* the professor covers that material in class. Thus, they are spending the *same amount of time* reading the chapter. They are simply doing it <u>one or two days too late</u> to be most effective.

Although no two people take notes the same way, the following suggestions might improve your note-taking skills.

Use large notebooks (8 1/2 x 11") instead of smaller, stenographer-type notebooks.

Leave a generous left margin to allow insertion of material, pictures, and diagrams. Use lined paper with a 3" left margin. If you cannot find this at an office supply store, try to find lined paper with no margin, and remember to leave yourself about 3" at the left.

Use a ballpoint or rollerball pen. If you make a mistake, just draw a line through it; you may discover later that you needed that information after all. Erasing takes time, and is irreversible.

Learn to outline. You cannot possibly write down everything your professor says, and it takes time to think and paraphrase the lecture. The only way to take complete notes is to write brief phrases and try to organize them logically. An outline is ideal for this.

Make important material stand out. Place an asterisk (*) next to important material, or underline it, or circle it. If you don't understand something during the lecture, make a note of it. It may help to write l.i.u. (= <u>l</u>ook <u>i</u>t <u>u</u>p) in the left margin when something is unclear during lecture.

Don't write in longhand. Although you may be accustomed to writing in attractive longhand, it takes much longer to make the curved lines, particularly on capitalized letters. Write just as fast as you can while still leaving something that is readable *to you.* Forget penmanship. Your finished product may be half-writing and half-printing, but as long as it is legible, that's fine.

Develop your own shorthand and abbreviations. There are many words and figures of speech that are common to many college courses. Whenever you can, use abbreviations for these phrases or words. (Think how much time you might save, for example, by just writing *hom* or *h.* instead of *homeostasis* each time the term comes up in lecture.)

Write as fast as necessary to ensure understandable notes. Write as much as you can during the period. Don't simply write what the professor puts on the chalkboard. It is easy to eliminate less-important material later. When you are taking notes, you don't know what is going to be important and what isn't. Write down everything you can while you have the chance.
Rewrite your notes soon after class. No kidding! Rewriting your notes takes little effort, and it is a wonderful form of mental reinforcement. Even if you are extremely conscientious in taking notes, there are going to be some points (typically *examples* of concepts) that you didn't have time to write during class. It is an excellent practice to spend some time rewriting your notes, preferably soon after the class. This will allow you to arrange the information logically, write more legibly, and add points that you remember from lecture but didn't have time to write. Students who tape the lectures have the added advantage of being able to rewrite their notes while listening to the lecture a second time. This may seem like too much trouble, but it is very effective. Rewriting notes doesn't take much time, the completed product is easier to study, and you can do it even if you are tired.

Is all of this really worth it? There are estimates that, for <u>each day </u>you spend in completing a college degree, you will earn an additional $1,000 during your lifetime. If you don't complete the degree, you can throw that all away; a partial college degree is worth the same as no degree.

Many "A" students are following the suggestions above, and don't seem a bit overtaxed by the process. You are competing with them for slots in nursing school, graduate school, and the job market.

If they are doing so much work, then why are *you* the one who is always feeling stressed out? Working efficiently is not stressful. Wasting your time with study habits that don't work is extremely stressful. None of the time you are putting in is working, and your failing grades make it *appear* that you aren't doing anything.

There is a Latin phrase, *quae nocent docent*, which means *things that hurt teach*. One good thing about mistakes is that we can <u>learn from them</u>. If your current study habits are NOT working, change them! Try some of these suggestions. You might miss some good television for now, but don't worry. Good television shows go into syndication; two years from now, they will be on *every day of the week*. If you are hooked on Oprah, Geraldo, Phil, Maury, and Sally Jesse, you may be seeking refuge in listening to people who are in worse shape than you. Turn *around* 180° and look at the world's successful people instead.

Cramming

Cramming is the term used to describe a long period of intensive studying just before an exam. Some books on "Making it in College" applaud the practice as the best way to get high grades with the least effort. They may be right, and if high grades with least work are all you are after, it might be right for you. Who would benefit from cramming?

Students who are taking classes whose content is irrelevant to their future endeavors.

Students whose grade point averages are so low that they must take "fluff" courses to raise it.

Unfortunately, this doesn't describe most students who cram. Students who cram are usually:

Those who procrastinate

Those with little ability to manage time

Those with little motivation or direction in their college studies

Those who fear that they won't be able to remember information longer than one day

Those who like to feel that they are "beating the system"

A few college study manuals advocate cramming with the following warning: they caution that cramming is most useful if you are not learning *new* information during the cram session. But, reviewing familiar material the night before an exam is not what most people call "cramming." People generally think of cramming as studying the bulk of the material for the *first time* the day, evening, or morning before an exam. As such, it is only useful as a last ditch effort to salvage an exam grade without resorting to cheating. You may get the grade you wanted, but you are unlikely to retain much of the material for long. In short, it is better than not studying at all. The fact that you must resort to it should alert you to your lack of effective time management.

The best way to study the night before an exam is to be so well-prepared that you don't have to do anything but quiz yourself and organize your thoughts. The priority you should have the night before an exam is getting seven to eight hours sleep. The only organ in your body that benefits from regular sleep is your brain. Staying up all night before an exam deprives your cerebrum of needed rest; it is the *only* organ that is going to get you through that exam successfully. Don't compromise your brain by overtaxing it the twenty-four hours prior to an exam.

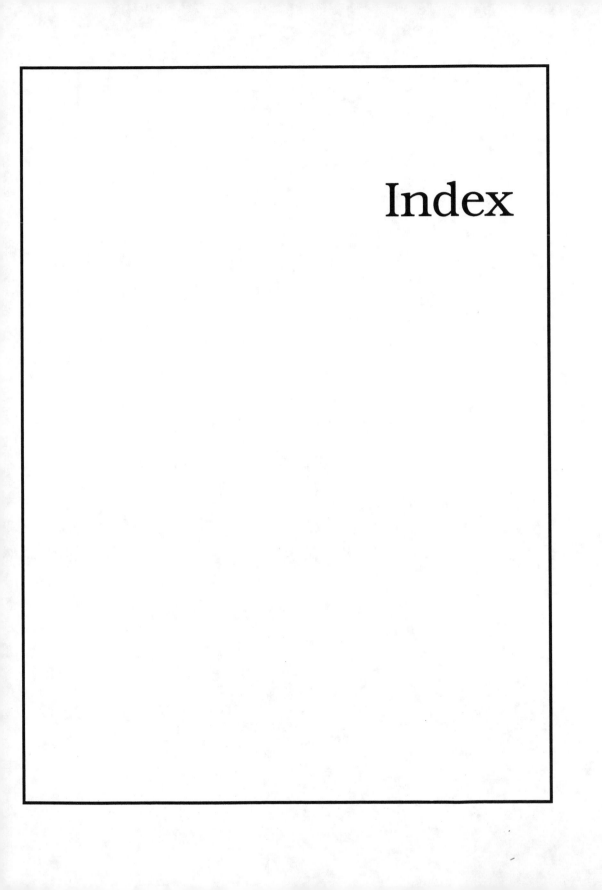

Index

A

Abdominal pregnancies, 149
Abdominal thrust, 94
Abnormal atrial or ventricular function, 68
Abrasions, 9
Abruptio placentae, 149
Abscess, 7
Absence seizures, 37
Accutane, 6
Acetazolomide, 129
Achalasia, 105
Achondroplasia, 11
Achondroplastic dwarf, 11
Acne, 6
Acquired immune deficiency syndrome (AIDS),
        82-88
Acromegaly, 12
Acupuncture, 43
Acute otitis media, 44
Acyclovir (Zovirax), 28, 144
Addison's disease, 53
Adult respiratory distress syndrome (ARDS), 98
AIDS-related complex (ARC), 83
Air Overexpansion Syndrome, 98
Alaxia, 30
Alcohol abuse, 121
Alcoholism, 121
Aldosteronism, 53, 134
Alopecia areata, 7
Alteplase, 72
Aminoaciduria, 128
Amnesia, 36
Anaphylactic shock, 77
Androgenital syndrome, 54
Anemia, 119
Aneurysm, 69
Angiotension-converting enzyme (ACE) inhibitors,
        73
Antabuse, 117
Anterograde amnesia, 40
Anthracosis, 96
Aortic stenosis, 62
Appendectomy, 74
Appendicitis, 74
Areflexia, 30
Artificial insemination, 144
Asbestosis, 91
Ataxia, 33
Atrial fibrillation, 68
Atrial flutter, 68
Audiogram, 45
Autodigestion, 114
Autologous marrow transplant, 56
Automatic bladder, 132
Avulsion, 9
Azidothymidine (AZT), 87

B

Bacterial vaginitis, 141
Baroreceptor accomodation, 68
Bed sores, 8
Bell's palsey, 35
Benzodiazepines, 41
Biliary obstruction, 115
Black lung disease, 91
Bleeding time, 58
Blood gas analysis, 96
Bone conduction, 46
Bone marrow transplant, 56
Borrelia burgdoferi, 88
Botulinus toxin, 19
Botulism, 19
Boyle's Law, 98
Brachial palsies, 30
Breech births, 152
Bronchitis, 95
Bronchography, 95
Bronchoscope, 94
Bronchoscopy, 94
Bumetanide, 129
Burkitt's lymphoma, 80

C

Calculi, 130
Calluses, 4
Candidiasis, 141
Captopril, 73
Carbon monoxide (CO), 102
Carbon monoxide poisoning, 102
Cardiac arrest, 68
Cardiac tamponade, 59, 77
Cardiogenic shock, 76
Cardiomyopathies, 64
Carditis, 59, 61
Carotene, 49
Castration, 137
Casts, 130
Cathartics, 110
Caudal anesthesia, 29
Cavities, 104
Cellulitis, 8
Cerebellar dysfunction, 30
Cerebral embolism, 71
Cerebral hemorrhages, 71
Cerebral meningitis, 29
Cerebral palsy, 36
Cerebral thrombosis, 71
Cervical cancer, 140
Chancre, 143
Chest tubes, 16

209

Epilepsies, 36
Epistaxis, 93
Epstein-Barr virus (EBV), 80, 82
Erysipelas, 8
Esophagitis, 105
Essential hypertension, 72
Estrogens, 138
Exophthalmos, 52
Exosurf, 98

F

Fertility drugs, 145
Fetal alcohol syndrome (FAS), 122, 151
Fibrinolysin, 8
Fibroids, 139
Fibromyalgia syndrome, 21
Filariasis, 78
First-degree heart block, 65
Flat feet, 18
Flexible flat feet, 18
Floaters, 50
Flurazepam, 41
Flutamide, 139
Focal seizure, 37
Forceps delivery, 149
Four-to-one block, 65
Fugu, 23
Furosemide, 129

G

Gallstones, 113
Gamete intrafallopian tube transfer (GIFT), 145
Gastrectomy, 106
Gastric stapling, 107
Gastric ulcer, 105
Gastritis, 105
Gastroenteritis, 108
Gastroscope, 106
Generalized seizure, 34
Genital herpes, 143
German measles, 89
Gestational neoplasm, 148
Giardia lamblia, 107
Giardiasis, 107
Glaucoma, 48
Glomerulonephritis, 124
Glycogen storage disease, 3
Goiter, 51
Gonorrhea, 142
Gout, 117
Gouty arthritis, 118
Graft versus host disease (GVH), 81
Grand mal seizure, 37

Graves' disease, 52
Guanidine hydrochloride, 20
Guillain-Barre syndrome, 27
Gynecomastia, 54
Gynelotrimin, 141

H

Halcion, 40
Hansen's disease, 31
Heartburn, 100
Heart block, 65
Heart failure, 62
Heavy metal poisoning, 26
Heimlich maneuver, 94
Helicobactor pylori, 105
Hematuria, 127
Hemolytic jaundice, 55
Hemophilia, 57
Hemorrhoids, 71
Hepatitis, 111
Hepatitis A, 111
Hepatitis B, 112
Hepatitis C, 112
Herpes virus, 143
Herpes zoster, 25
Heterologous marrow transplant, 56
Heterotopic bones, 13
High ceiling diuretics, 129
Hirsutism, 7
Hives, 6
Hodgkin's disease (HD), 79, 93
Horner's syndrome, 42
Human immunodeficiency virus (HIV), 82
Human papilloma virus (HPV), 140
Huntington's disease, 39
HV-1, 142
HV-2, 143
Hyaline membrane disease (HMD), 98
Hyaluronidase, 8
Hydatidiform mole, 148
Hyernatremia, 127
Hypercalcemia, 60
Hyperkalemia, 60, 134
Hyperkeratosis, 5
Hypernatremia, 132
Hyperostosis, 12
Hyperparathyroidism, 53
Hyperthyroidism, 52
Hyperrefexia, 32
Hypersomnia, 41
Hypersplenism, 82
Hypertension, 72
Hyperthyroidism, 52
Hyperuricemia, 117
Hyperventilation, 103
Hypervitaminosis, 115

Nephritis, 127
Nephrolithiasis, 130
Neurogenic shock, 77
Neurosyphilis, 143
Neurotoxins, 25
Night blindness, 49
Nodular, 75
Nomogram, 135
Non-Hodgkin's lymphoma (NHL), 79
Nonunion, 13
Nosebleeds, 93
Nutrasweet, 116

O

Obesity, 118
Obstructive jaundice, 55, 114
Obstructive shock, 77
Oophorectomy, 145
Open wound, 9
Opportunistic infections, 83
Orchiectomy, 137
Orthostasis, 73
Orthostatic hypotension, 73
Osmotic diuretics, 129
Osteogenesis imperfecta, 11
Osteopetrosis, 12
Otitis media, 44
Ouabain, 72
Overhydration, 132

P

Palsies, 30
Pancreatic obstruction, 115
Pancreatitis, 114, 115
Pap smear, 140
Para-aminohippuric acid (PAH), 125
Parainfluenza viruses, 93
Paralytic shellfish poisoning (PSP), 25
Parasomnias, 41
Paresthesia, 30
Paroxysmal atrial tachycardia (PAT), 68
Partial placenta previa, 149
Partial thromboplastin time (PTT), 58
Pelvic inflammatory disease (PID), 142
Peptic ulcer, 105
Perforated ulcer, 106
Pericarditis, 59
Periodontal disease, 104
Peripheral neuropathies, 30
Peroneal palsy, 31
Petit mal epileptic attack, 37
PET scans, 38
Phenylketone, 116

Phenylketonuria (PKU), 115
Pheochromocytoma, 54
Phrenology, 15
Pinkeye, 42
Placenta previa, 149
Plaque, 104
Plasma prothrombin time (PT), 58
Pleural effusion, 92
Pleurisy, 97
Pleuritis, 97
Pneumocystis carinii, 83, 95
Pneumonia, 95
Polio, 22
Poliovirus, 22
Polycystic kidney disease, 125
Positive end-expiratory pressure (PEEP), 98
Post-traumatic amnesia (PTA), 40
Potassium sparing diuretics, 129
Preeclampsia, 151
Pre-emetic phase, 108
Pregnancy induced hypertension (PIH), 152
Premature atrial contractions (PACs), 68
Premature ventricular contractions (PVCs), 68
Premenstrual syndrome (PMS), 144
Preparatory phase, 108
Pressure palsies, 30
Prevention of AIDS, 86
Primary hypertension, 72
Primary syphilis, 143
Prochlorperazine (Compazine), 45
Promethazine, 45
Prostatectomy, 138
Prostate-specific antigen, 139
Prostatic acid phosphetase, 139
Prostatitis, 138
Protein deficiency diseases, 116
Proteinuria, 127
Pseudohermaphrodite, 147
Psoriasis, 5
Pulmonary edema, 74
Pulpitis, 104
Punctures, 9
Pus, 8
Pyelitis, 130

Q

Quinacrine, 107

R

Rabies, 24
Radial nerve palsy, 31
Red tide, 25
Regulatory obesity, 118

Photo Credits

CHAPTER 7 **7A** John Radcliffe Hospital/Science Photo **7B** ©1974 L.B. Halstead Wykeman Publications (London) Ltd.

CHAPTER 10 **10A** March of Dimes Birth Defects Foundation

CHAPTER 14 **14A** Custom Medical Stock Photo

CHAPTER 15 **15A** Courtesy Dr. Michael M. Ter-Pogossian, Washington University School of Medicine **15B1** Courtesy Dr. Michael M. Ter-Pogossian, Washington University School of Medicine **15B** National Institute on Aging

CHAPTER 17 **17A** ©1990 Meyer: Custom Medical Stock Photo **17B** Visuals Unlimited/(c) SIU

CHAPTER 20 **20A** Biophoto Associates/Photo Researchers **20A** (top) SIU/Custom Medical Stock Photo **20A** (bottom) SIU/Custom Medical Stock Photo

CHAPTER 22 **22B** Centers for Disease Control **22C** NASA **22D** Ken Greer/Visuals Unlimited

CHAPTER 23 **23A** AP/Wide World Photos

CHAPTER 24 **24A** Biophoto Associates/Photo Researchers

CHAPTER 25 **25A** United Nations

CHAPTER 29 **29A** Lester V. Bergman & Associates